International Conference on Power and Propulsion Alternatives for Ships 2019

London, United Kingdom
23 January 2019

ISBN: 978-1-5108-9541-6

Printed from e-media with permission by:

Curran Associates, Inc.
57 Morehouse Lane
Red Hook, NY 12571

Some format issues inherent in the e-media version may also appear in this print version.

The Royal Institution of Naval Architects
8-9 Northumberland Street
London, WC2N 5DA
United Kingdom

Phone: 020 7235 4622
Fax: 020 7259 5912

publications@rina.org.uk

Additional copies of this publication are available from:

Curran Associates, Inc.
57 Morehouse Lane
Red Hook, NY 12571 USA
Phone: 845-758-0400
Fax: 845-758-2633
Email: curran@proceedings.com
Web: www.proceedings.com

Power & Propulsion Alternatives for Ships

23 January 2019, London, UK

PAPERS

THE ROYAL INSTITUTION OF NAVAL ARCHITECTS
8-9 Northumberland Street, London, WC2N 5DA, Telephone: +44 (0)20 7235 4622

The Royal Institution of
Naval Architects

INTERNATIONAL CONFERENCE

Power & Propulsion Alternatives for Ships

23 January 2019, London, UK

Papers published in these conference Proceedings are peer reviewed.

THE ROYAL INSTITUTION OF NAVAL ARCHITECTS
8-9 Northumberland Street
London WC2N 5DA

Telephone: 020 7235 4622
Fax: 020 7259 5912

CONTENTS

1 **Alternative fuels: The Present and Future of Containment Systems and Their Impact on the Design and Construction of Ships**
Fabrizio Cadenaro, Ed Fort, Liam Blackmore, Lloyd's Register EMEA, UK

9 **New Shuttle Tanker Concept**
Jon Nation, Jeroen Van Keep, Kjell Storelid, Wärtsilä Marine Solutions, UK

12 **Modelling Alternative Propulsion Technologies for Merchant Vessels**
John Buckingham, David Pearson, BMT, UK

18 **Can Biofuels Help Provide Clean Propulsion for Shipping, Now and in the Future?**
Chester Lewis, Dr Ausilio Bauen, E4tech Ltd, UK

23 **Examining Methanol as an Alternative Marine Fuel for Indonesian Domestic Ships**
Eko Maja Priyanto, Biro Klasifikasi Indonesia, Indonesia
Aykut I Ölçer, Dimitrios Dalaklis, Fabio Ballini, World Maritime University, Sweden

39 **Towards Electrification of Ro-Ro Passenger Fleet in the Adriatic Sea**
Maja Perčić, Ivica Ančić, Nikola Vladimir, University of Zagreb, Croatia

45 **How to Reach Total Power Train Efficiency**
Juho Rekola, Mika Koli, Steerprop Ltd, Finland
Martin Andtfolk, SE Tech, Finland

52 **Comparison of Diesel-Electric with Hybrid-Electric Propulsion System Safety Using System-Theoretic Process Analysis**
Victor Bolbot, Gerasimos Theotokatos, Evangelos Boulougouris, Dracos Vassalos, University of Strathclyde, UK

59 **Agile Power Management Systems – A Rule-Based Control Strategy Using Real-Time Simulation for Hybrid Marine Power Plants**
Chris Watts, Babcock International Group, UK
Dr Truong Quang Dinh, Dr Truong Minh Ngoc Bui and Dr James Marco, Warwick Manufacturing Group (WMG), University of Warwick, UK

68 **Investigation of Auxiliary Power Potentials of Solar Photovoltaic Applications on Dry Bulk Carrier Ships**
Wandifa Saidyleigh, Gambia Maritime Administration
Aykut I Ölçer, Raphael Baumler, World Maritime University, Sweden

83 **Modern Rotor Sail Technology Helps Ships Save Fuel and Reduce Emissions - Performance and Experiences from Recent Installations**

Ville Paakkari, Norsepower Oy Ltd, Finland

88 **Case Study: Wind-Assisted Ship Propulsion Performance Prediction, Routing, and Economic Modelling**

Nico van der Kolk, Giovanni Bordogna, Paul Desprairies, TUDelft, the Netherlands

James Mason, University of Manchester, UK

ALTERNATIVE FUELS: THE PRESENT AND FUTURE OF CONTAINMENT SYSTEMS AND THEIR IMPACT ON THE DESIGN AND CONSTRUCTION OF SHIPS

F. Cadenaro, **E. Fort**, **L. Blackmore**, Lloyds Register EMEA, UK

SUMMARY

In recent years, driven primarily by SOx emissions regulations, there has been a move towards the adoption of Liquefied Natural Gas (LNG) as a marine fuel. More recent decarbonisation targets, and the emissions regulations that are due to follow, will almost certainly trigger a further move towards other low carbon, carbon neutral and zero carbon fuels. Methanol, ammonia, and hydrogen offer a potential pathway for the decarbonisation of the shipping industry. However, the various physical and thermodynamic properties of such alternative fuels will require new containment systems onboard ships requiring the marine industry to embrace containment technologies used in other industries or, where necessary, develop industry-specific solutions. Shipboard containment systems for diesel fuels are currently physical, based on storage at near ambient temperatures and ambient pressures and for natural gas at high pressure (compressed) or low temperature (cryogenic), or a combination of such. Future containment technologies are likely to also include material and chemical based storage, exploiting chemical processes to hydrogenate or dehydrogenate carriers, in both liquid or solid matrices. This paper provides an overview of alternative fuels and their containment technologies and the implications on ship design and construction.

1. INTRODUCTION

Like most, if not all industries, shipping is ultimately shaped by public perception and public opinion. Fuels that are perceived to be safe and clean are likely to be viewed more favourable.

The IMO has acted decisively on short-term and long-term regulations affecting emissions from shipping. This is enacted through 2020 Global Sulphur Cap, IMO Initial Strategy for GHG Reduction from International Shipping, EEDI and IMO DCS. These are not discussed in this paper even though they provide the motivation.

Until now natural gas (methane), in its liquefied form, liquified natural gas (LNG), has been the most developed and utilised alternative fuel in the industry. It was first used as fuel during the carriage of LNG as cargo on board Gas Carriers, due to their inability to prevent the boil-off phenomenon, in which a fraction of the cryogenic cargo evaporates due to heat leakage into the cargo containment system. The early consuming systems were steam boilers which supplied superheated steam to a turbine power plant, whereas now the current practice is to consume it directly in dual fuel diesel engines.

The use of cargo boil-off provides a solution for a specific industry sector whereas, the rest of the industry will need to view the future fuels more holistically.

Societal and recently marine industry aspirations are towards zero carbon fuels. The safety hazards, physical and thermodynamic properties and containment system architectures associated with such alternative fuels are well known and their use is commonplace in many industries. However, their use onboard ships of all types is novel and potentially challenging. Before widespread adoption can occur, such fuels must be adequately understood. Collaborative projects between industrial partners provide a robust manner to achieve the required understanding.

This paper considers the containment and storage of a number of alternative, low carbon, carbon-neutral and zero carbon fuels currently considered most likely to figure in the future fuel mix, in particular natural gas (CH_4), ethane ($C2H_6$), methanol (CH_3OH), ammonia (NH_3) and hydrogen (H_2). Storage of these fuels is typically achieved physically using pressurised or cryogenic containment or a combination of both. Other storage techniques using materials capable of absorbing or adsorbing gases such as hydrogen in a liquid or a solid matrix with the potential for increased safety particularly when used onboard ships. Some of these solutions assist with the volatility of the fuel while achieving increased volumetric densities. The required storage technologies to achieve acceptable volumetric densities, in some cases, depart radically from traditional ship design culture and knowledge.

2. FUEL PROPERTIES AND THEIR IMPLICATION ON THE DESIGN AND CONSTRUCTION

2.1 SUMMARY OF FUEL PROPERTIES

Table 1 summarises the properties of the key fuels with reference to diesel oil. This table is referred to and quoted in the Chapters contained in Section 2.

2.2 NATURAL GAS (METHANE)

Natural gas is a mixture of molecules where the bulk is methane, which constitutes between 80% to 98% of the volume. Other hydrocarbons are present, such as ethane and propane, these are generally heavier but exhibit similar behaviour, while the other fractions are carbon dioxide, nitrogen, helium and hydrogen sulphide that exist in small percentages.

The overall physical and chemical behaviour of natural gas is similar to pure methane, where the boiling point at atmospheric pressure is approximately -163°C. Its critical point is approximately -83°C at 46 bar. This implies that natural gas needs to be stored in liquid form at cryogenic temperatures or as compressed gas (CNG) at ambient temperature to achieve an effective energy density. LNG has a density ranging from 430 kg/m^3 to 480 kg/m^3, depending upon its composition, which is more than 600 times the gas density at room temperature. With reference to Table 2, it is noted that LNG has a density advantage over CNG, even when the latter is compressed to hundreds of bars.

From a ship design and construction perspective, the energy demands of international shipping is encouraging the development of alternative fuel installations to reach volumetric energy densities as close to the current installations as possible. The on board bunkered energy capacity for fuel oil can exceed 100TJ. Despite that the volumetric energy density of both CNG and LNG being significantly less than existing fuel oils, the use of LNG has become an established marine fuel where competent shipyards, operators and a bunkering infrastructure exist.

From a regulatory perspective the use of natural gas as a marine fuel is fully addressed with Goals, Functional Requirements and Prescriptive Requirements.

2.3 ETHANE

Ethane is an organic chemical compound with the formula C_2H_6. At ambient temperature and pressure, it is an odourless and colourless gas. Its properties are similar to methane but has a higher boiling point and density. All other thermodynamic properties such as the flammability limits are similar.

As with methane, the detection and ventilation of the gas is a key requirement to achieve regulatory approval. However, unlike methane, ethane is neutrally buoyant in air, as such, the designer needs to demonstrate adequate safety systems.

Ethane is a common by-product of petroleum refining and natural gas processing. It is primarily used to produce ethylene. In recent years, the growth of this market segment has provided greater market opportunities for its transportation and use as cargo as fuel.

At present, there is no prescriptive regulatory landscape for ethane as a marine fuel. However, this does not prevent its use – to use ethane as fuel or ethane cargo as fuel would require risk studies to demonstrate compliance as necessitated by the SOLAS Convention and applicable Codes.

2.4 METHANOL

Methanol, also known as methyl-alcohol, with a chemical formula CH_3OH - often abbreviated MeOH. Methanol can be produced both by distilling organic products or from hydrocarbons through an industrial catalytic process. The motor racing industry has used methanol as a fuel due to its knocking resistance and smoke free combustion. It is chemically and physically similar to ethanol. At room temperature, it is a transparent liquid with a boiling point at 65°C and a density of 0.79 kg/l at atmospheric pressure. It is rather volatile and miscible with water and unlike ethanol it is toxic. Its lower heating value (LHV) is approximately 19.9 MJ/kg less than half the typical value of diesel fuel [2]. Methanol vapours have a large flammable range in air, with LEL 6% and a UEL 36% by volume [1]. Methanol vapours are 10% heavier than air at ambient temperature and pressure.

From a ship design and construction perspective, the thermodynamic properties most relevant are the LHV and density. The LHV indicates that for a given energy storage capacity, the bunkered volume required by methanol is twice that of diesel fuel, if consumption, fuel weight and efficiencies are considered constant, which they are not. Only considering this, there are clear implications for passenger numbers and/or cargo capacity. The vapour density of methanol also presents additional design considerations for detection. In the event of a methanol leakage, the vapours may descend and accumulate in spaces and compartments in the ship's hull rather than being exhausted, either; naturally or by a forced ventilation system. This compounded by the low boiling point of the fuel requires additional precautions in the handling, bunkering and storage. An alternative detection and ventilation system architecture is required to prevent and reduce the associated risks.

From a regulatory perspective, the use of methanol as a marine fuel is addressed in the high-level Functional Requirements of the IGF Code. Compliance would be achieved using risk studies. The development of the supporting Functional Requirements and Prescriptive Requirements is in progress at the IMO, where the Sub-Committee on Carriage of Cargoes and Containers has completed the drafting of the methanol and ethanol fuel guidelines.

2.5 AMMONIA

Ammonia is an inorganic compound of nitrogen and hydrogen with the chemical formula NH_3. Ammonia is used as a fertilizer, where regional overproduction is common, as such, it is a widely transported commodity.

The critical temperature of ammonia is 132°C with a boiling point of -33°C. It is therefore carried at a low temperature in isolated tanks or at room temperature in pressurised containers. The molecular packing structure of ammonia means that more hydrogen is contained per unit volume than liquid hydrogen. This indicates that ammonia could be a credible hydrogen fuel or carrier with an easier transition.

As a fuel, it can be consumed by Internal Combustion Engines (ICE) and has the future potential to be consumed by fuel cells. The slow flame speed of ammonia means that an unstable combustion process occurs. This can be overcome by enriching the supply fuel with hydrogen through reforming a percentage of the ammonia.

Anhydrous ammonia is transported on board LPG vessels and as bulk on board Containerships. The cargo is carried in compliance with the IGC Code and IBC Code respectively.

2.6 HYDROGEN

Hydrogen is the most abundant chemical element in our universe with the symbol H and atomic number 1. It typically bonds to form the H_2 molecule. Most of the hydrogen on earth is bonded with oxygen in the form of water, H_2O. This implies that unless obtained by fossil fuels such as methane through a reforming process, it is not a primary source of energy, but an energy vector, meaning energy must be spent to produce it, which will be recovered when oxidised. As such hydrogen has the potential to be generated from renewable energy – this would produce a so-called green vector if produced from water, whereas, fossil fuels produce a brown vector fuel.
Hydrogen at atmospheric temperature and pressure is a light, odourless, colourless and a non-toxic gas. It is highly flammable and easily forms explosive mixtures with air. Its liquefaction temperature is approximately -253°C at atmospheric pressure and its critical point is approximately -240°C at 13 bar [4]. It is, therefore, necessary to cryogenically cool hydrogen to liquefy it and gain a meaningful energy density. Its liquid density is abt. 70 kg/m3, less than one-tenth of a typical hydrocarbon fuel, whereas its LHV is 120 MJ/kg which is three times more than diesel fuel or natural gas.
From a processing perspective, it is important to note that hydrogen occurs in two isomeric forms: ortho and para. These states are differentiated by their proton spin where para-hydrogen is a lower energy state than ortho-hydrogen meaning that as hydrogen is cryogenically cooled the molecules will transition. This contributes to the vapourisation phase transition [19]. As the isomeric form is associated with boil-off, then the energy differential should be accommodated in the containment design calculations.
From a ship design and construction perspective, liquefied hydrogen has a poor volumetric energy density. The vessel architecture would need to consider hazardous zones, autoignition characteristics, liquid to gas expansion ratio, cryogenic temperatures, flammable range, and potential for liquid releases to produce dense vapour clouds. Whereas, the system design needs to accommodate the permeability and embrittlement behaviours. The current safety concepts as described in the IGF Code; Gas Safe and Emergency Shutdown (ESD) need to be especially considered during the application of hydrogen. Some of the characteristics assist designers in the philosophy: Prevention, Detection and Ventilation, such as the gaseous buoyancy, however, many do not.
The use of hydrogen from a regulatory perspective can be addressed by the Goals and Functional Requirements defined in the Codes. As previously described for other fuels, no Prescriptive Requirements exist. Hydrogen has been identified as a zero-carbon fuel by the IMO in their Initial Strategy for Reduction of GHG Emissions from Ships (MEPC.72) meaning a regulatory framework is on the horizon for the IGF Code. It should be noted, that guidelines already exist for the carriage of liquefied hydrogen as a cargo in Resolution MSC.420(97). Lloyd's Register recognises the need for further understanding in the use of hydrogen cargo as fuel and hydrogen as a bunkered fuel. The Society is currently engaged with governments, academia and industry to achieve this greater breadth and depth of knowledge.

3. FUEL CONTAINMENT SYSTEMS AND THEIR IMPLICATION FOR SHIP DESIGN AND CONSTRUCTION

3.1 PHYSICAL CONTAINMENT

3.1 (a) Structural tanks

Structural tanks are the simplest and most common fuel containment system on board ships. They are spaces within the hull where boundaries are created by decks and bulkheads. They achieve a 98% space utilisation including the filling limit, however, due to their basic construction, there is an inability to withstand a pressure differential. It is possible for these tanks to be modified for the carriage

	Diesel	Natural Gas	Ethane	Ammonia	Methanol	Hydrogen
Boiling temperature [°C]	180+	-163	-88	-33	65	-255
Liquid density [kg/m^3]	840	430	542	682	790	70.8
Gas density [kg/m^3] @ 20°C	-	0.67	1.25	0.7	-	0.084
Flame temperature in air [°C]	2100	2230	2220	1850	1950	2396
Heat of vaporisation [J/g]	233	510.4	498	1365	1200	449
Lower flammability limit [% vol.]	0.6	5	3	15	6.7	4
Upper flammability limit [% vol.]	7.5	15	12.4	28	36	75
Minimum ignition energy [mJ]	0.1	0.28	0.26	680	0.14	0.017
Flash point [°C]	> 60	-188	-135	11	12	< -253
Auto-ignition temp. [°C]	315	537	472	651	433	585
Toxicity	Low	No	No	Yes	Yes	No
Temperature at critical point [°C]	-	-83	32	132	239	-240
Pressure at critical point [bara]	-	45.9	49	112.8	78.5	12.9

Table 1: Properties of fuels (various sources)

of alternative fuels that are liquid at ambient temperature and pressure; such as, methanol and ammonia. The designer should have due regard for material compatibility, material longevity, and where necessary, inerting of the tanks vapour space. Design regulations for the storage of low flashpoint fuels, lower Design regulations for the storage of low flashpoint fuels, lower than 60 °C, do not currently exist. A design would be derived from a risk studies that would demonstrate compliance with the IMO IGF Code Goals and Functional Requirements.

3.1 (b) Prismatic IMO Type A tanks

IMO Type A tanks, as defined by the IGC and IGF Codes, are non-pressurised independent prismatic tanks designed using ship structural analysis procedures and recognised standards. They are primarily constructed of plane surfaces and the design pressure is no more than 700 mbar. Where cold gases below -10°C at near to atmospheric pressure exist, a full secondary barrier is required. A secondary barrier is an additional containment barrier that can contain the worst envisaged leakage while protecting ship's hull from contact with cryogenic spills. Type A tanks offer high space utilisation, simplicity of construction, and their design and construction is regulated by the IGC Code, IGF Code and Lloyd's Register's Rules and Regulations.

3.1 (c) Prismatic or spherical IMO Type B tanks

IMO Type B tanks are independent of the ship's hull. These tanks are designed using analytical tools, model tests and direct calculation methodologies, to determine stress levels, crack propagation and fatigue life of the tanks. They are not considered to be pressure vessels and can form various shapes, including prismatic, while the most known implementation is the Kvaerner-Moss type spherical design. Since the tank structural behaviour is fully analysed in the design process, only a partial secondary barrier is required to safely contain potential leakages. The space utilisation is high for prismatic tanks, while poor for spherical, although spherical tanks are the most widely used on LNG Carriers. This tank design is well defined in the IGC and IGF Codes.

3.1 (d) Pressure vessels, IMO Type C tanks

IMO Type C tanks are independent of the ship's hull. The tanks are designed and built according to pressure vessel standards which include both fracture mechanics and crack propagation criteria. There is no upper limit to the design pressure; however, most tanks are approved for pressures between 5 to 10 barg. They are typically designed with a cylindrical shape and in some instances with a bi-lobe shape for improved space utilisation. For large tanks and bi-lobe tanks, the thermal insulation is made with polyurethane foam or similar, applied to the outer surface, whereas smaller cylindrical tanks are built using the "double wall" technique. Double wall describes incorporating one tank inside another, creating an annular space, which is kept under vacuum and filled with perlite or multilayer insulation. This is a more complex solution that provides efficient thermal insulation but may provide a secondary barrier. In general, Type C tanks are not required to have a secondary barrier; however, this may be challenged when a tank has one or more pipe penetrations below the maximum liquid level. This situation does not exist for Gas Carriers, where suitable domes for piping connections and instruments can be accommodated. It can be problematic for other vessels where sufficient height below deck restricts the penetration design. In this case, the requirement for the secondary barrier around the tank may be fulfilled by an outer vessel. This would then be considered as double walled.

This tank type is a common choice for natural gas vessels, due to: Administrations accepting their presence below deck with the appropriate risk studies, their capability to withstand pressure, and ability to store gas with little to zero boil-off. The design is robust - it can handle sloshing, is cost effective, and the design has reached maturity. This tank design provides an opportunity with the appropriate research, development, and risk studies to store hydrogen in a cryo-compressed phase [5][6].

The design and construction of Type C tanks are fully regulated by the IGC and IGF Codes, and LR's Rules and Regulations.

3.1 (e) Membrane tanks

A membrane tank is composed of two gas-tight membranes. The primary and secondary barriers are made of stainless steel or other suitable metals. A layer of thermally insulating material is fitted between the membranes and to the surface of the outer barrier. All the forces including the pressure, the weight of the liquid gas, and sloshing are transmitted to the ship's hull. The insulation space is inerted with nitrogen, displacing the air, to reduce the risk of fire (in case of leakage) and to avoid the condensation of moisture.

This tank type is not self-supporting and is built as part of the hull, meaning that the ship's structure and containment system are dependent on each other, requiring strong communication between the shipyard and the containment system designer. There are several containment systems of this type available - the most common of which are:

- GTT No 96 technology is based on Invar membranes where insulating layers are built up with plywood boxes. These boxes are filled with perlite and are secured in place with couplers and tie-rods;
- GTT Mark III technology is based on a primary membrane made of stainless steel with corrugations to accommodate thermal expansion. Insulating layers are constructed of load bearing prefabricated panels in reinforced polyurethane foam. The secondary

barrier is made of Triplex, a composite laminated material [10].

Other variants and iterations of the above-mentioned systems exist. The membrane system achieves excellent space utilisation and has high form adaptability. They are in general complex to construct due to the integration with the ship's hull. These tanks are designed to carry liquefied gases at near-atmospheric pressure, making the management of the boil-off gas necessary, which unlike Type C tanks, cannot be managed by pressure accumulation. This is likely to limit their implementation for non-cargo carrying vessels.
Regulatory requirements for membrane tanks can be found in the IGC and IGF Code, and in LR's Rules and Regulations.

3.1 (f) Lattice tanks

Cylindrical and spherical tanks have a form that reduces the hoop-stresses produced by the pressure differential across the tank boundary. However, their space utilisation is poor. Lattice style tanks overcome this by form-fitting the space and implementing internal structural members to support the pressure surfaces. The pressure vessel could be theoretically built to accommodate any space while still managing boil-off with pressure accumulation. In addition to this, their construction is simplified when compared to membrane tanks. At present these tanks have only been demonstrated with a single-skin design, this is unlikely to be acceptable for enclosed spaces, which limits their advantage for form-fitting.
This tank design is considered to be novel, as such, no marine regulatory regimes exist. These tanks would need to be constructed under an Alternative Design and Arrangements scheme to achieve compliance to the IGC or IGF Code.

3.2 ALTERNATIVE CONTAINMENT SYSTEMS

3.2 (a) Metal Hydrides

Metal hydrides are metals which can bond hydrogen to the material elements. This material containment system achieves a greater volumetric density of hydrogen than compressing the molecule in a gaseous phase. The US Department of Energy has set goals for energy density and volumetric energy density which would need to be achieved for the technology to be competitive in the automotive industry, refer to Figure 1. The estimated values indicate that the containment system could store more hydrogen than is achievable in both a gaseous compressed and liquefied phase.
To extract the fuel, the hydride is heated which liberates the hydrogen from the covalent or ionic bond. Therefore, the heat input is effectively subtracted from the efficiency of the system and becomes a function of the tank size [8]. This process of absorption and release is called hydrogenation and dehydrogenation respectively.

The most promising metal hydrides, in terms of storage capacity, are those working at a temperature above 150°C. These require significantly less energy for dehydrogenation [12]. Supply and return pressures affect the performance of the containment system, where the dehydrogenation pressure is required to be high enough for the consumers, otherwise additional energy is required for compressors. This is currently under research in the development of metal hydrides [17].

3.2 (b) Liquid Organic Hydrogen Carriers (LOHC)

LOHC describes a hydrocarbon medium that can cyclically absorb and release hydrogen by means of a catalytic reaction. These hydrocarbons are not themselves the fuel but act as a carrier. An example of this is Dibenzyltoluene. This LOHC can acquire 18 hydrogen atoms during the hydrogenation process or 9H2 molecules - it then becomes known as Perhydro–Dibenzyltoluene [11]. The amount of hydrogen that can be stored is expressed as a percentage of the substance weight, where currently the technology can achieve up to 6%. The carrier mediums have a similar density to water, this translates to a hydrogen storage capacity of 60 kg/m3. This is comparable to liquid hydrogen at 71 kg/m3 [9].
Once the LOHC substance has been "depleted" to a hydrogen poor condition - the spent LOHC should be unloaded and hydrogenated LOHC bunkered. The spent LOHC would then be enriched at a handling facility. The advantage of this technology is that it negates the requirement to cryogenically cool or compress the hydrogen. Some carriers are considered non-toxic; however, this does not extend to all. The bunkering, handling, storage and containment designs and processes should reflect the associated hazards. The transportation, handling, and distribution of LOHC could utilise the existing land-based and on board infrastructure, including the piping and tanks, while considering material compatibility.

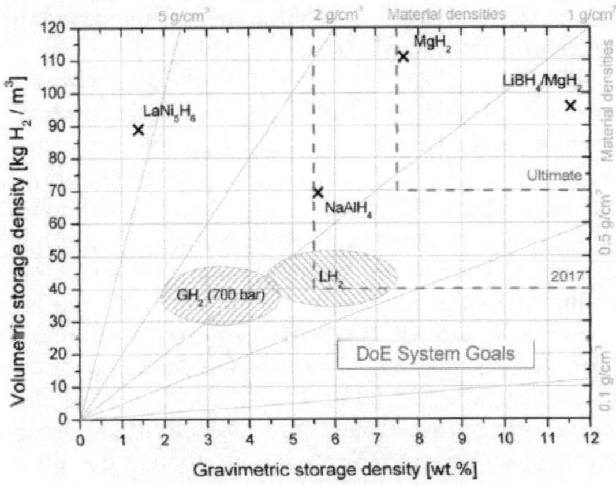

Figure 1: Gravimetric and volumetric density of H2 storage technologies. Courtesy of US Department of Energy

3.2 (c) Ammonia as a hydrogen carrier

As previously stated, ammonia contains more hydrogen per unit volume than liquefied or compressed hydrogen. Ammonia as a carrier can liberate its hydrogen through a reformer, where the reformer will realise between 50-70% of the hydrogen. However, complete ammonia reforming is unnecessary for an ICE. Approximately 10% hydrogen is required to stabilise the combustion process due to the slow flame speed of ammonia. Alternatively, when hydrogen is intended to be used in fuel cells – 100% of the ammonia carrier is required to be reformed. This combined with the potential for ammonia contamination presents a challenge. Residual ammonia in the supply fuel may adversely affect the fuel cells operation and the longevity of the cells.
The production of nitrous oxides during the reforming and combustion of ammonia is unavoidable. The exhaust streams can be treated with Selective Catalytic Reduction. This reduces the NOx molecules into water vapour plus molecular nitrogen using a chemical process.

3.2 (d) Class Approach

At present, there are no Classification rules or regulations available for hydride technologies. These are still considered novel, as such, the Alternative Design and Arrangements (Risk Based Design Process) would be required. The process is scalable according to the degree of novelty, design complexity and safety considerations. Hence, the time and effort for each stage will vary. In short, the process comprises of the following stages: Stage 1 – Design and Safety Statement, Stage 2 – Risk Assessment, Stage 3 – Revision and Supporting Studies, and Stage 4 – Final Design Assessment.

4. STORAGE EFFICIENCY

The move to alternative fuels is happening now. Comparing alternative fuels to MGO and HFO provides a known datum to reference but this also highlights the unattractive characteristics of these fuels while negating their environmental credentials.
A containment system is composed of many interacting components that store, deliver, and convert an energy source into something consumable by energy converters (e.g. engines and fuel cells). The space utilisation, volumetric density, thermodynamic storage conditions, thermodynamic properties and ultimately energy converter efficiency, all determine the fuel and total containment system efficiency. In this paper, we discuss the energy density of the fuels with only a reference towards the absolute system attributes.
To simplify this, and make the results more tangible, Table 2 represents estimates and characteristics of the key efficiency drivers. The space utilisation is based on experience and the LOHC figures are estimated on future predictions by the US Department of Energy [5].
Traditional carbon-based fuels are an efficient means for energy storage, see Table 2. By comparison, methanol achieves less than half the energy density, while LNG varies between a half to a third depending on the storage system. However, these storage technologies are driven by the ship type and therefore the economics. Type C tanks offer one-third of the energy density but substantially reduce the amount of on board equipment required to control the boil-off, whereas membrane tanks provide the economics of scale to justify reliquification plants etc. Methanol has a comparable storage efficiency to LNG while also having similar environmental credentials. The advantage of methanol would be the reduced CAPEX.
The highest hydrogen energy densities are achieved by complex containment systems and supporting infrastructure; this is demonstrated by liquefied hydrogen (-253°C and atmospheric pressure). The combination of the heating value, energy density, and storage efficiency computes to an equivalent tank size greater than four times, while the equivalent bunkered mass would be substantially lower.
Ammonia can be used as a fuel or a hydrogen-carrier where as a carrier it would suggest that 100% of the ammonia is reformed on board. In fact, ammonia is likely to be bunkered for both. The inefficiencies of reforming would demand additional storage capacity which is unlikely to be economical. However, 10-20% reforming for flame stabilisation is justifiable. And as research and development progresses on ammonia fuel cells, the reforming, along with the addition tank capacity will diminish.

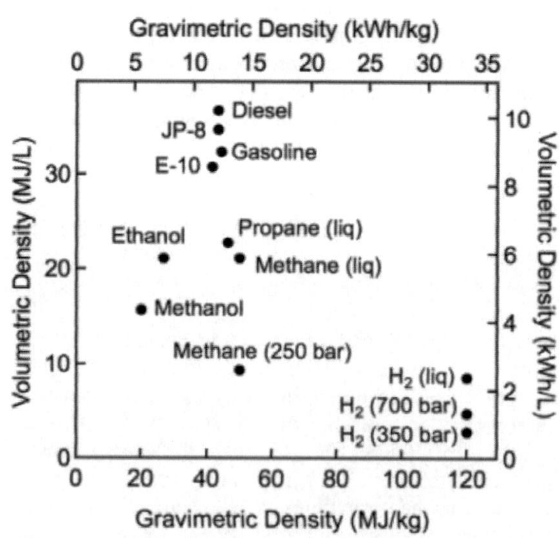

Figure 2: Courtesy of US Department of Energy – Energy Efficiency & Renewable Energy

Fuel	Storage	Bulk density [kg/m^3]	Energy density [GJ/m^3]	LHV [MJ/kg]	Space utilisation
Diesel fuel	Structural tank	840	36.1	43	~ 100%
HFO	Structural tank	980	39.2	40	~ 100%
Methanol	Structural tank	790	15.7	19.9	~ 100%
LNG	Membrane	430	19.6	48	~ 95%
LNG	Type C	450	10.3	48	~ 50%
CNG	Press. 200 bar	180	4.3	48	~ 60%
CNG	Press. 700 bar	370	8.9	48	~ 50%
LH2	Type C double wall	71	4.3	120	~ 50%
CH2	Press. 200 bar	18	1.1	120	~ 50%
CH2	Press. 800 bar	67	4	120	~ 50%
LOHC	Structural tank	~ 1000	7.2	120	~ 6% H_2 wt.
NH_3 (for H_2)	Type C	682	7.2	120	~ 50%

Table 2: Energy density of different fuels and storage technologies

5. CONCLUSIONS

Containment technologies for a number of alternative fuels have been reviewed. Current containment systems for the traditional marine fuels like diesel oils and heavy fuel oils, based on physical containment technology at near ambient temperatures and pressures, have proved relatively straight forward in respect of design and construction and adequate in respect of energy storage efficiency. Ship designs and bunkering schedules have evolved accordingly.

Emerging marine fuels such as liquefied or compressed natural gas also utilise physical containment but at elevated pressures or cryogenic temperatures or a combination of both. Consequently, such containment systems are significantly more complex and, due to the lower volumetric energy density of liquefied or compressed natural gas coupled with the reduced energy storage efficiency of pressurised and cryogenic storage technology, offer a reduced energy storage capacity when compared to fuel oils. However, the increasing use of LNG as a marine fuel suggest such a reduction, although undesirable, can be accommodated with current ship designs.

Unfortunately, other potential marine fuels have still lower volumetric energy densities. Methanol has a lower volumetric energy density compared to liquefied natural gas but physical storage at near ambient conditions results in an overall energy storage capacity not dissimilar to that of LNG but much more straight forward design and construction requirements. Hydrogen, the ultimate zero carbon fuel, has even lower energy density, both compressed and liquefied and consequently alternatives to physical storage are currently under investigation within the industry, in particular the potential of material and chemical based storage. The use of organic hydrides (LOHC) materials may offer a comparatively safe and effective storage material for hydrogen while the use of chemicals such as ammonia as the store or carrier of hydrogen offers the prospect of significantly increased energy storage capacity compared to compressed or liquefied hydrogen storage.

The major driver for the adoption of low carbon, carbon neutral and zero carbon fuels is the reduction, and ideally elimination of harmful emissions. However, such aspirations appear to come with a cost; in terms of, the complexity of the containment systems, fuel expenditure due to reforming, system management, and the increased physical storage space necessary compared to traditional fuel oils with its potential impact on the cargo or passenger carrying capacity of existing ship designs operating existing trading routes.

6. REFERENCES

[1] https://www.mathesongas.com/pdfs/products/Lower-(LEL)-&-Upper-(UEL)-Explosive-Limits-.pdf

[2] https://www.engineeringtoolbox.com/fuels-higher-calorific-values-d_169.html

[3] https://en.wikipedia.org/wiki/Natural_gas

[4] IMO MSC.420(97)

[5] US department of energy, Technical Assessment: Cryo-Compressed Hydrogen Storage for Vehicular Applications, Revised June 2008

[6] Sofoklis S. Makridi, Department of Mechanical Engineering, University of Western Macedonia, Greece - Hydrogen storage and compression.

[7] J. Jepsen, Technical And Economic Evaluation of Hydrogen Storage Systems based on Light Metal Hydrides, Helmholtz Zentrum Geesthacht, 2014

[8] https://energy.gov/eere/fuelcells/hydrogen-storage

[9] Liquid Organic Hydrogen Carriers (LOHC): An auspicious alternative to conventional hydrogen storage technologies - J. von Wild, T. Friedrich, A. Cooper, B.

Toseland, G. Muraro,W. TeGrotenhuis, Y.Wang, P. Humble, A. Karim

[10] http://www.gtt.fr/en/technologies-services/our-technologies

[11] Hydrogen Storage in Liquid Organic Hydrogen Carriers – insights from first project experiences - Hydrogenious Technolgies, Hannover, April 2017

[12] HYDROGEN STORAGE AND FUEL PROCESSING STRATEGIES, Callum Richard Campbell, 2014.

[13] International Code of the Construction and Equipment of Ships Carrying Liquefied Gases in Bulk (IMO)

[14] International Code of Safety for Ships using Gases or other Low-flashpoint Fuels (IMO)

[15] LR Rules and Regulations for the Classification of Ships using Gases or other Low-flashpoint Fuels

[16] LR Provisional Rules for the Classification of Methanol Fuelled Ships

[17] EU FCH Programme Review 2015

[18] Monterey Gardiner. 2019. DOE Hydrogen and Fuel Cells Program Record. [ONLINE] Available at: https://www.hydrogen.energy.gov/pdfs/9013.

[19] Ronald M. Bliesner, RB, JULY 2013, 'PARAHYDROGEN-ORTHOHYDROGEN CONVERSION FOR BOIL-OFF REDUCTION FROM SPACE STAGE FUEL SYSTEMS', WASHINGTON STATE UNIVERSITY, USA.

7. AUTHORS BIOGRAPHY

Fabrizio Cadenaro holds the current position of Gas Fuelled Ships Senior Specialist at Lloyds Register EMEA. He is based in Trieste, Italy and is responsible for gas-as-fuel technical coordination within LR. His previous experience includes many years in the design, engineering and appraisal of gas carriers, gas fuelled ships and relevant storage, piping and safety system.

Ed Fort currently holds the position of Global Head of Engineering Systems at Lloyds Register. He is responsible for Engineering technical policy within Marine & Offshore. Originally a seagoing engineer his previous experience includes hydrogen fuel cell engineering in both marine and aerospace applications.

Liam Blackmore currently holds the position of Systems Engineering Specialist at Lloyd's Register in the Global Technical Centre in Southampton UK. He has previously held permanent positions in Europe, Canada and China with LR.

NEW SHUTTLE TANKER CONCEPT

J Nation, **J V Keep**, **K Storelid**, Wärtsilä Marine Solutions, UK

1. INTRODUCTION

It is raining innovations in Wärtsilä. The new Shuttle Tanker concept allows tankers to operate both on Liquefied Natural Gas (LNG), as the primary fuel, and Volatile Organic Compounds (VOC) – which is the environmentally harmful gas evaporating from the oil cargo tanks – as secondary fuel.

The new innovative solution aims to reduce annual emissions of CO_2 equivalents by up to 42% compared to conventional shuttle tanker. The reduction equals the emissions from 22,000 cars per vessel per year.

The main task for a shuttle tanker is to transport oil from offshore fields to land-based oil terminals. As shuttle tankers operate in offshore conditions, these vessels are the most advanced merchant vessels with unique varieties in operation modes including loading in Dynamic Positioning (DP) condition, transit in laden, unloading and ballast conditions.

All of these operating modes have their own requirements, but traditionally these requirements have led to equipment installed on board not being used efficiently in the various operation modes.

The new Shuttle Tanker will have the following three features compared with a conventional shuttle tanker:

- **Reduced emissions**: VOC emissions from the cargo will be eliminated, the NOx from the engine exhaust will be reduced by 84% which is well below IMO Tier 3 levels, while the SOx emissions will be practically eliminated, and finally the particles will be reduced by more than 96%, thus resulting in an astonishing reduction of emissions.

- **Operation flexibilities**: an efficient use of the installed machinery and propulsion systems in the vessel secures an unmatched manoeuvring capability while the built-in system redundancies ensure inherent system robustness when managing unexpected events.

- **Improved economics**: 22% reduction in total fuel consumption and with the use VOC as fuel, resulting in considerable reduction in bunkering needs, and combined with fewer running hours hence lower maintenance costs for machinery.

2. REDUCED EMISSIONS

During operation, conventional shuttle tanker design has high levels of emissions, mainly from the emissions of VOC to the atmosphere during loading, storage, and transportation of crude oil. For instance, the storage and loading of crude oil onto ships are responsible for more than 50% of Norway's VOC emissions. The Norwegian authorities have thus implemented stringent emission reduction regulations for all shuttle tankers loading crude oil from offshore processing platforms located in the Norwegian continental shelf.

The loading operations take place when crude oil is loaded to the Shuttle Tanker directly via a loading hose connected to a Floating Production Storage Offshore (FPSO) unit, fixed platform or via a loading buoy. This operation can take place in harsh environments and with difficult sea stages. The crude oil need to be loaded with high transfer speed to ensure predictability in weather conditions throughout the DP operation.

During the offshore crude oil loading, the VOCs are emerging from the crude oil cargo tanks and must be captured by a VOC recovery plant, avoiding harmful emissions to the atmosphere.

Wärtsilä has long experience in developing VOC recovery plants that can satisfy the requirements from the Norwegian authorities. With these experiences, Wärtsilä has now designed a new generation of VOC recovery plant that will satisfy also the more stringent authority requirements expected from 2030.

The Wärtsilä VOC recovery plant uses compression and cooling phases to liquefy the heavier hydrocarbons to Liquid VOC (LVOC) that is stored in a tank on the deck of the vessel.

The lighter hydrocarbons that are not liquefied, referred to as Surplus VOC (SVOC), which mainly consists of methane gas, will be burnt in a gas turbine for electricity generation, chosen because of the two times better efficiency than the traditional use of boiler with steam generator.

From a typical North Sea platform installation, each crude oil loading of a cargo size of 850,000 bbls, will recover 100 tonnes of LVOC and 10 tonnes of SVOC. By avoiding these amounts of VOC escaping into the atmosphere, the yearly reduction of emission goes from 43,000 to 25,000 tonnes of CO2, i.e., a reduction of 42%, assuming around 32 loadings per year.

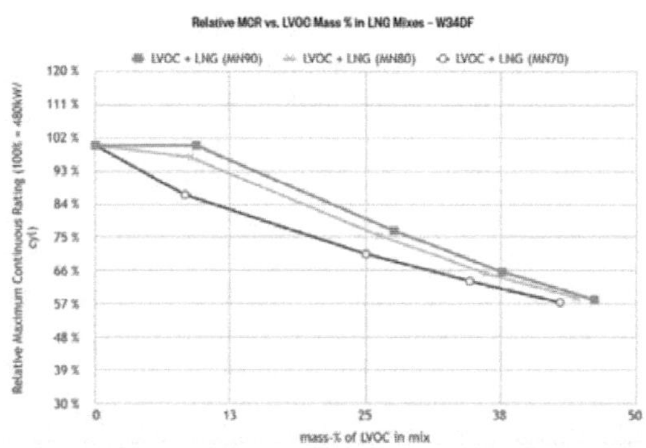

Fig. 1 - Relative MCR vs. LVOC Mass % in LNG Mixes – W34DF

3. USE OF LNG AND LVOC AS FUEL FOR ENGINE

Today's shuttle tankers are equipped with direct propelled 2-stroke diesel engines running on heavy fuel oil (HFO) or marine gas oil (MGO) and the recovered LVOC has so far been considered as a waste product. The 2-stroke engines are mainly used for propulsion during transit, and the 4-stroke auxiliary plant is providing power for the extensive thruster system used during DP operation.

With an average of 100 tonnes of recovered LVOC per loading, the recovered amount could represent up to 30% the total fuel consumption of the Shuttle Tanker should we be able to use the LVOC as fuel.

TEEKAY has therefore replaced the conventional 2- and 4-stroke engine configuration with fully electric propulsion with generator sets driven by Wärtsilä 4-stroke dual-fuel (DF) engines.

With electric main propulsion motors and 4-stroke DF generating sets as the only power plant onboard the vessel, flexibility and overlapping functionality is achieved. This power distribution concept is part of Wärtsilä Low Loss Hybrid concept reducing the total installed power on board from 26 to 23 MW with further reductions in fuel consumption and resulting in an increased overall efficiency of the vessel.

To achieve full SECA and NECA compliance, the new Shuttle Tanker will be further equipped with Wärtsilä LNGPac to enable operation in gas mode throughout the shuttle tanker operation with LNG as primary fuel for the engine.

With both LNG and LVOC onboard, Wärtsilä started to develop and performed testing of the possibility of mixing LNG with LVOC in gas form for potential valuable fuel for the engine.

LVOC comprises heavier gas hydrocarbons such as propane and butane that have relatively low methane number (MN) of 25, making them less suitable as fuel for our gas engines (to avoid knocking). By mixing the LVOC with LNG with an MN between 70 and 90, we will achieve an acceptable MN for the gas engine at any required power. Depending on the MN of LNG the engines will run a maximum continuous rating (MCR) less than 100% due to the de-rating mode. This de-rating mode is fully acceptable due to the operational flexibility built into the new Shuttle Tanker concept. (Figure 1)

Utilising LNG as the primary fuel and LVOC as secondary fuel, the new Shuttle Tanker enables the utilisation of 100% of the recovered LVOC as fuel for electric power generation. (Figure 2)

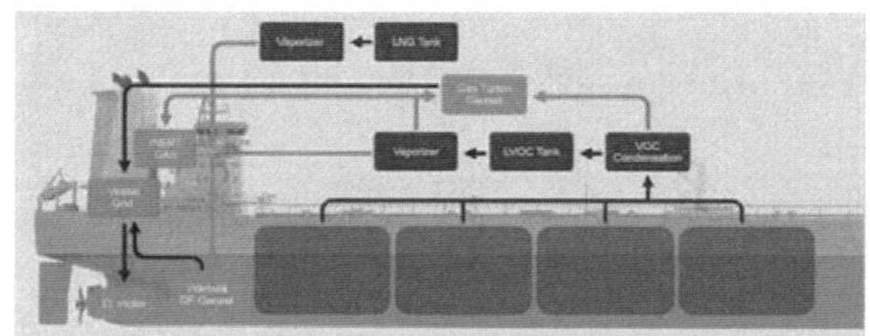

Fig. 2 - With an average fuel consumption of 8750 tonnes per year, 3200 tonnes will be generated and used onboard as LVOC.

4. OPERATIONAL FLEXIBILITY

The power distribution system will be based on the Wärtsilä Low Loss Hybrid (LLH) system using batteries for further fuel savings, peak load shaving and added overall system redundancy.

Wärtsilä LLH does not only offer the highest efficiency over all electrical components, but a great benefit comes from its superior capability to minimise the impact of a failure during DP operation. In addition to the required trial speed of the vessel, the size and functionality of the power plant is determined from the power requirement in DP condition. While a traditional electrical distribution system could lose more than 50% of installed power and several thrusters, the LLH will lose only 25% of the installed power and not more than one thruster.

Thanks to the LLH system, the system efficiency is improved and together with electric propulsion, the required total mechanical installed power will be reduced by 14%, which will improve overall fuel performance compared with traditional power distribution concepts.

As a result of the new Shuttle Tanker concept, the total energy consumption will go from 110 GWh to 75 GWh per year compared to a traditional shuttle tanker.

With increased efficiency, we will also improve the DP (manoeuvring) capability. A conventional system will use 60% of thruster power, whereas in LLC concept only

40% will be used, thus giving the master a much better manoeuvring capability.

Moreover, the LLH electric equipment room is more compact compared to a conventional electric room. The Shuttle Tanker power distribution system utilises a Low Voltage arrangement also with excess of 23 MW of installed power, which will again have a large benefit for the crew on board.

When installing batteries on board, the vessel has a hybrid energy system that will continuously support the dynamic capacity of the energy production system and, through the stored energy of the batteries, enable engines to operate in a load area where fuel consumption is optimal. The batteries will handle the dynamic load variations and the engines will get a stable load and therefore could operate in a higher load area without risking the startup of additional generators due to transient load variations. The New Shuttle Tanker is the first ship of this size utilising batteries for improving efficiency during transit operation.

For crude oil offloading operations at the onshore terminals, Wärtsilä can supply either electric-driven pumps for pump room installation, or electric-driven deep well cargo and ballast pumps that enable a distributed pump solution, eliminating the need for a separate pump room and interconnecting pipelines in the cargo holds.

The space gained from eliminating the pump room can be used either to increase the cargo capacity or to shorten the engine room with a shorter hull, giving lower building cost and better DP capability thanks to the leaner side profile.

With the efficient and environmentally friendly electric power plant installed on board the vessel, electrical-driven pumps for cargo off loading and ballast are utilised in the most efficient way.

In connection with ballast pumps, Wärtsilä can supply its Ballast Water Management System (BWMS) named AQUARIUS® EC. It uses an in-situ chlorine generation technology known as electro-chlorination (EC). AQUARIUS® EC is a modularly designed BWMS. The system is installed in a bypass loop to the main ballast line to provide a safe, flexible and economical process for the treatment of ballast water and eradication of aquatic invasive species.

5. IMPROVED ECONOMICS

The new Shuttle Tanker concept represents an absolute game-changer for the shuttle tanker sector. To put it in simple terms, the concept enables shuttle tankers to travel from the oil fields on its own waste gas rather than releasing it in the atmosphere. Both emissions and bunkering requirements will be considerably reduced and the chosen propulsion solution will increase overall efficiency compared with conventional systems.

With this concept, the overall running hours will be lower, which will mean lower maintenance costs for machinery.

Wärtsilä is also ready to support and provide active assistance during the operation of the new Shuttle Tanker vessels, to accomplish a total life-cycle solution to the owner. Such support and assistance will comprise both training of the crew and the officers for the equipment and the systems together with varied technical and operational real-time online support and monitoring. Such operational follow-up directly with the owner is crucial to ensure that the vessel achieves the objectives set for the new Shuttle Tanker in respect to emission reduction, fuel efficiency and with operational flexibility.

These benefits have been made possible with Wärtsilä's advanced technology and through integration capabilities between different business segments in Wärtsilä supplying the various equipment and systems.

Fig. 3 - The new Shuttle Tanker is the first ship of this size utilising batteries for improving efficiency.

6. MARKET POTENTIAL

TEEKAY has contracted two firm and two options of the new Shuttle Tanker vessel at Samsung in Korea, all based on Wärtsilä's high-quality equipment and systems in this new and innovative concept solution.

We estimate that half of all offshore oil production in Norway and the UK will be transported to land using shuttle tankers. About 40% of this existing tanker fleet is ready for replacement due to the 20-year age limitation for new shuttle tankers.

We see an exciting time ahead for similar contracts in the near future for this ecological and economical Shuttle Tanker concept developed in cooperation with TEEKAY and Wärtsilä.

MODELLING ALTERNATIVE PROPULSION TECHNOLOGIES FOR MERCHANT VESSELS

J E Buckingham and D R Pearson, BMT Defence and Security UK Ltd

SUMMARY

The planned IMO sulphur emissions limit will likely lead to price increases of low-sulphur distillate fuels from 2020 and beyond, and drive further pressure to improve fuel efficiency. The IMO measures planned for 2023 to meet the 2050 IMO CO_2 reduction target will formalise the need for improved fuel efficiency through Energy Saving Technologies (EST).

A wide variety of EST have been available for many years: the challenge lies in integrating and matching them to the ship's operating profile to provide demonstrable benefit, and thereby improve adoption rates. With many ships operating in slow-steaming mode, there is opportunity for wind-based devices to contribute a larger proportion of the required thrust. This in turn provides scope to explore how reduced engine loads affect fuel efficiency, the use of air lubrication drag reduction and the use of wild heat from engine jacket water and exhaust gases to generate power.

As part of a collaborative project with partners Black and Veatch, BMT has been developing the Vessel Technology Assessment System (VTAS) funded by the Energy Technologies Institute (ETI). This project has developed an approach that seeks to demonstrate the merit of ESTs fitted to specific ships on specific voyages.

This paper presents the approach taken to use mathematical models of the ship's power and propulsion system to identify the fuel saving benefits of a set of EST option designs. The performance before and after the installation of ESTs is modelled and the results compared to the baseline design.

NOMENCLATURE

ALDR	Air lubrication drag reduction
CFD	Computational Fluid Dynamics
DG	Diesel Generator
DP	Design Point
DWT	Deadweight Tonne
EEDI	Energy Efficiency Design Index
EST	Energy Saving Technology
ETI	Energy Technologies Institute
FR	Flettner Rotor
HFO	Heavy Fuel Oil
IMO	International Maritime Organisation
MRV	Monitoring, Reporting, Verification
ORC	Organic Rankine Cycle
P&P	Power & Propulsion
PTI	Power take-in
PTO	Power take-off
Sfc	Specific Fuel Consumption (kg kWh^{-1})
STW	Speed Through Water
SWH	Significant Wave Height
TG	Turbo-Generator
TPC	Tonnes per Centimetre
VTAS	Vessel Technology Assessment System
WHR	Waste Heat Recovery

1. INTRODUCTION

1.1 OBJECTIVE

To assess the benefit of any single EST or EST combination on a given ship, it is first necessary to build ship power & propulsion (P&P) models which are aligned with the ship's actual performance as captured through ship-board sensors and then recorded as ship-data.

Such an alignment for a series of data points then provides a firm baseline for the effect of an EST to be assessed against within the modelling space.

1.2 STUDY SHIP

This study considers a 61,000 DWT bulker whose principal particulars are given below:

- Length overall 199.9m
- Length waterline 197.0m
- Beam 32.24m
- Draught 13.04m

Such a ship definition is given at the Design Point (DP), the point where the working displacement and hence the draught is greatest. In service the ship will be more lightly loaded with a lower draught and probably out of trim. Using the Holtrop & Mennen standard propulsion analysis method [1] such deviations from the DP are accommodated so that the actual ship can be faithfully modelled.

The paper outlines the methods used to model actual ships at sea in real world sailing conditions so that the benefit of EST in such operating scenarios can be faithfully identified with a confidence related to the likely margin of error. A more detailed explanation of the modelling approach has previously been described [2].

This provides more insight into engine and environmental modelling.

2. SHIP DATA

2.1 SENSORS

The increasing use of sensors to capture the key operating parameters of machinery allows this data to be used for more than the usual health and fouling assessment.
The most useful data is listed below:

Navigational	Machinery	Meteorological
Vessel location	Shaft revolutions per minute (rpm)	Wind Speed & Direction
Time & Date	Shaft torque & therefore power	Significant Wave Height (SWH)
Speed & course over Ground	Main Engine fuel mass flow	Current Speed & Direction
Heading	Auxiliary engine fuel mass flow	Air temperature & humidity
Speed Through Water (STW)	Draught fwd & aft	Water temperature

Although ship data, usually aggregated to 1 or 5 min periods, has inconsistencies due to sensor accuracy and the rounding effect over the sample period, it represents the best data set to use for these studies as it is the closest possible to the complete definition of a ship's behaviour at sea.

3. MODELLING

3.1 APPROACH

Whether there is already ship data available or whether the study is to start with the most basic of data, the ship's propulsion system is to be modelled to a level consistent with the quality of results required for decision making.
The ship and its machinery is modelled using the ship's principal particulars, with the environmental data used to derive the following:

- Ship's resistance
- Propeller speed
- Engine loading
- Fuel demand

Each of these stages requires the best possible definition of the hullform, the propeller and the engine. If working in close co-operation with the ship owner many of these parameters can be provided. However, when performing individual generic ship studies the full data set is not available.

Examples of unknown hydrodynamic parameters are the wake fraction and thrust deduction factors. Ideally they are taken from a model test report or CFD study. If such sources are not available, then they are derived from a statistical analysis of a range of empirical prediction methods as shown in Figure 1. With experience, those models which are best suited to specific ship types and sizes can be identified. Figure 1 shows a spread of estimates for the thrust deduction factor using the mean; an awareness of the confidence value at a given margin of error allows an appreciation of the accuracy to be established.

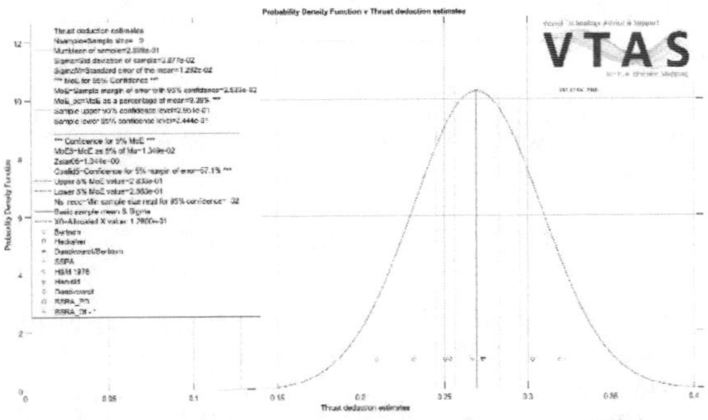

Figure 1. Thrust deduction estimates against probability density function

Our recent experience is that for thrust deduction and wake fraction estimation, these empirical methods are increasingly becoming less accurate (thrust deduction is usually much lower) and wherever possible CFD data for comparable ships should be sought.

Although the propeller Kt-Kq characteristic can sometimes be supplied, where this is not possible, the propeller behaviour is estimated using the Wageningen B-series [3].

3.2 VOYAGE ANALYSIS

In order to compare the vessel's power & propulsion with and without EST with appropriate confidence, it is necessary to have good models for both the ship and the EST. The nature of such models is now explained.

3.3 ENGINE DEFINITION

Measured shaft power and speed correlated with the main engine fuel flow data allows a Specific Fuel Consumption (Sfc) map of the engine to be generated, which can then be used with the ship model to characterise fuel use in different conditions. However, when such data is not available the engine manufacturers Sfc data with a ±5% tolerance is used as shown in Figure 2.

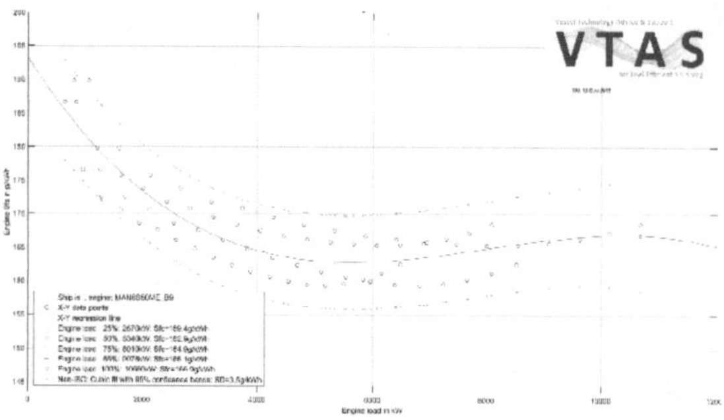

Figure 2. Engine Sfc v Load from Engine Supplier's information

4. SHIP MODEL DEVELOPMENT

The ship definition is based on the principal particulars which with the hullform coefficients which can be used by Holtrop & Mennen [1] to provide a ship resistance across the speed range. Estimates of added resistance in waves is provided by using the Alexandersson method.

The development of the modelling approach relies upon the estimation of key parameters such as wake fraction and thrust deduction factor where model test reports are not available. When the ship draught is away from the DP, as often is the case, the tonne per centimetre (TPC) factor is used to determine the actual displacement.

This set of hydro-dynamic coefficients and factors is used to determine the best estimate of resistance in this condition. Where appropriate an allowance for hull roughness & fouling is applied.

5. ENERGY SAVING TECHNOLOGY MODELS

The energy saving technologies considered as part of this study generally fall into 3 categories, euphemistically referred to as "wet, warm and windy":

Wet: Hydrodynamic efficiency technologies, primarily consisting of specialist hull paints, propeller pre/post swirl devices, and other water-side technologies such as air lubrication drag reduction.
Warm: Refers to Waste Heat Recovery (WHR) technologies such as Organic Rankine Cycles (ORC), and Turbo-Generator (TG), utilising heat from jacket water or exhaust gas.
Windy: Wind-based auxiliary propulsion such as Wingsails, Flettner rotors, Kites or other types of sailing method.

5.1 APPROACH

The approach presented in Figure 3 is to use air lubrication drag reduction (ALDR) in the form of a layer of microbubbles to reduce the ship's principle hull resistance.

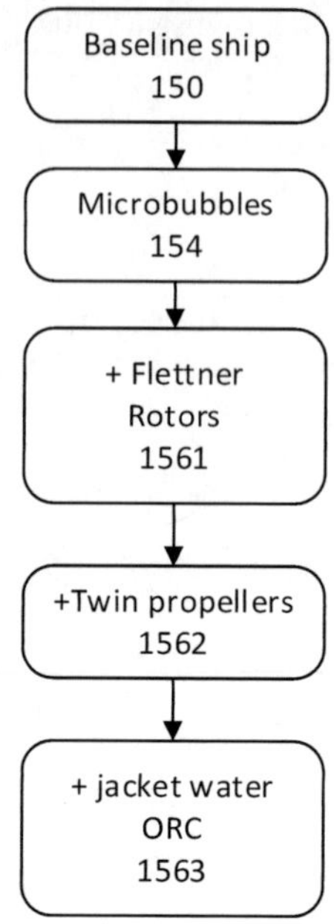

Figure 3. Study Option Flowchart

5.2 AIR LUBRICATION DRAG REDUCTION

Air Lubrication Drag Reduction (ALDR) uses a thin layer of microbubbles in the boundary layer to reduce the viscous friction coefficient at the water-hull interface. The size of the bubble is critical as large bubbles can deform and shear.

Moriguchi & Kato [4] used bubbles between 0.5mm and 2.5mm and measured up to a 40% decrease in resistance, for an air content proportion of over 10%.

The provision of air at the pressure matching the water at the ship's keel is to be produced as efficiently as possible. These studies use the rule of thumb of 0.12kWh/Nm3 for an acceptable specific energy consumption [5]. Mizokami et al. [6] conducted ALDR testing on a full-size mock-up and then on a 162m bulker of 19,000 tonnes displacement. The ship-test involved thee different equivalent air thickness sizes of 3, 5 and 7mm with the beneficial propeller shaft power reduction for a given speed being as shown in the table.

Parameter	Values		
Equivalent air thickness in mm	3	5	7
net energy-saving effect in %	8	10	12

The net benefit is derived from the reduction in gross propulsion power required and hence fuel saved, less the compressor power consumed. Mäkiharju [7] uses a first principles polytropic equation to determine the compressor power required but a practical approach is to use the compressor supplier data. Compressor efficiency varies from 90% for good quality reciprocating designs through to 60% for turbo-compressors.

Design issues

Kawakita [8] identifies potential issue of the ingress of air into sea chests and how this may have to be actively managed. Kawakita also reviews the potential issues for adverse propeller fluctuations due to void fraction distribution but no firm conclusions are offered.

Inappropriate use of ALDR can also lead to an increase in resistance and differing results between modelling and full-scale experience [9]. Lay et al. [10] observed that the formation of an air layer can be significantly hindered by any random perturbations on the surface upstream of the gas injector. Clearly then detailed design is a key feature of an ALDR system.

Project

The ALDR model employed here is described by Kodama, [11] & [12]. Our modelling assumes a compressor efficiency of 80%.

5.3 WIND-BASED EST

Wind-based EST benefit from higher sea-state conditions where windier conditions prevail. Although the benefits are influenced by the relative wind direction and the potential for significant added resistance in waves, the benefits of reducing the ship's sea-going resistance at source leads to lower propeller loading with a better efficiency, whether it be via ALDR, Flettner Rotors or Wingsails.

5.4 FLETTNER MODEL

The Flettner Rotor (FR) model is based on previous work performed in this area [13]. As Flettner rotors are a wind-based device, the benefit provided by them is highly dependent on the environmental conditions experienced by the vessel. The model determines the useful forward FR thrust for a range of sea states and wind headings. For a typical general purpose study, sea state 4 is used as a good reference for the time-weighted average of the whole range of sea states. The thrust benefit for the whole azimuth is averaged to determine the benefit for that sea state.

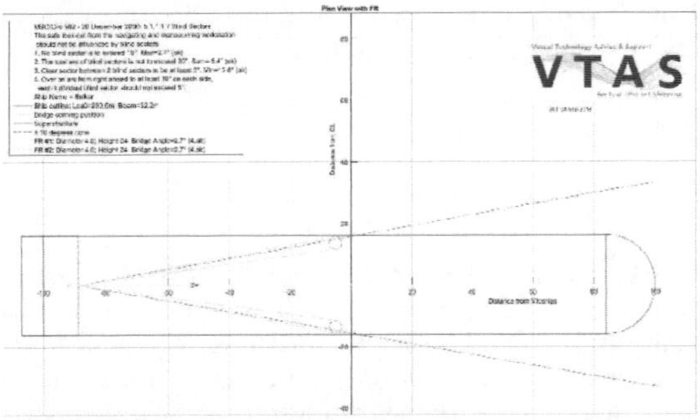

Figure 4. FR location: Plan View

Figure 4 shows a plan view of the location of the two FR on the ship's deck. The positions are automatically checked to ensure the FR position are valid iaw IMO regulations and do not obscure the visibility from the ship's Bridge

The FR model operates under a set of operational rules that simulate its behaviour on a real vessel. These include restricting operation unless it achieves a net propulsive benefit, i.e. the equivalent propulsive power exceeds the rotor power requirements and any generated drag.

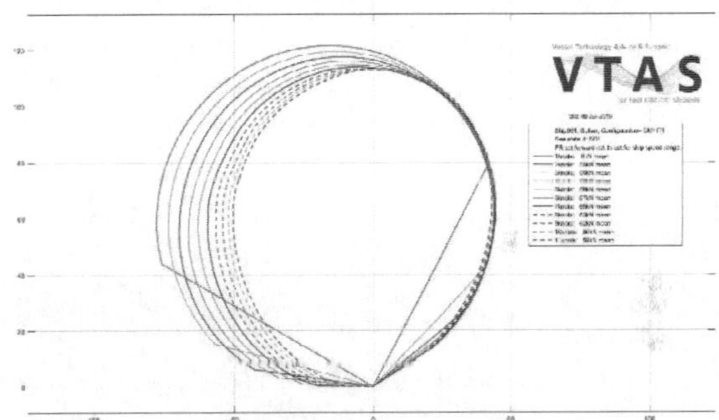

Figure 5. FR Forward thrust v True Wind Angle

Figure 5 shows how the FR benefits vary with true wind angle with the best performance being slightly aft of the beam. The assessment takes the average of the set of full azimuth thrust to provide a balanced assessment of the FR benefit.

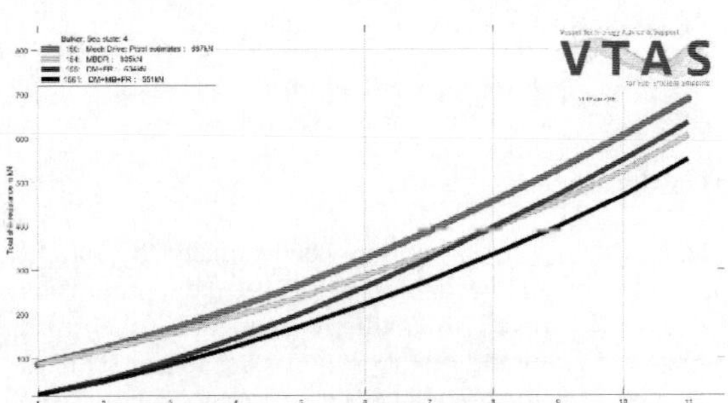

Figure 6. Resistance with ALDR & FR EST

Figure 6 shows how the combination of FR and ALDR are complimentary (shown in black versus a baseline of red), leading to a reduced overall resistance across the speed range.

5.5 TWIN SHAFT PROPULSION

The baseline single fixed pitch propeller driven by a slow speed two-stroke main engine is a common propulsion arrangement. The whole propulsion thrust is provided via the limited area of the single propeller. This leads to a higher thrust coefficient and a lower theoretical propeller efficiency. A twin propeller arrangement as employed by the Maersk Triple-E class offers a theoretical improvement of ~20% with no propeller diameter change. The reduced propeller loading leads to slower shaft speeds which allows for recent changes in engine tuning to be fully introduced.

Project

The baseline design has a MAN 6S50ME engine rated at 8,130 kW, 108 rpm driving a single 6.25m diameter propeller. The twin shaft variant has two 6.25m propellers, each driven by a MAN5S40ME two-stroke engine rated at 4,325 kWb. Although the power range is sufficient for a four stroke to be used with a gearbox and PTO-PTI options, a two-stoke design is retained in this solution to retain the basic P&P architecture.

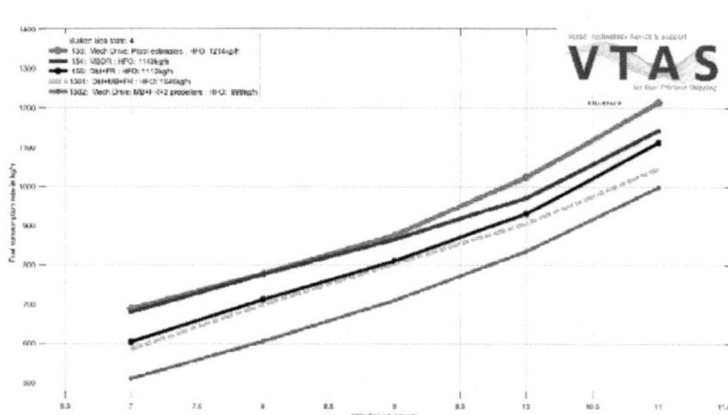

Figure 7. Fuel Consumption with ALDR, FR & twin shafts

The red line in Figure 7 shows how a twin shaft design offers a significant fuel saving benefit.

5.6 WASTE HEAT RECOVERY

Technology

After the vessel resistance has been reduced by ALDR and wind-based technologies and the propulsion efficiency improved by multiple propellers, the main engines are required to deliver the power to the shaft. A waste heat recovery (WHR) system recovers energy from the jacket water using Organic Rankine Cycles (ORC) and from the exhaust gas using advanced ORC or steam. Power generated in this way allows the auxiliary diesel generator (DG) set to operate at lower loads. With exhaust gas ORC the ship's electrical power demand may be fully supplied from the main engine with excess power potentially being utilised as part of a Power-Take-In (PTI) arrangement to drive the ship. However, the main engine usually has an economiser which transfers exhaust gas waste heat to thermal oil to keep the HFO warm. As a minimum temperature may be required for an SCR to operate to keep NOx emissions below the IMO MARPOL Annex VI limits, an exhaust gas WHR solution has not been considered in this study.

5.7 ORGANIC RANKINE CYCLE

One of the most common ORC at sea for the recovery of jacket water heat is the Calnetix Hydrocurrent ORC [14], rated at 125kWe. Smaller feasible units based on a commercial land-based unit from AVID or Electratherm have been defined to provide ~35kWe per MAN5S40ME main engine.

Project

For this project, the ORC would take jacket water heat at 90°C and leave enough heat in the jacket water (75°C) after the ORC for it to drive a vapour distillation flash evaporator.

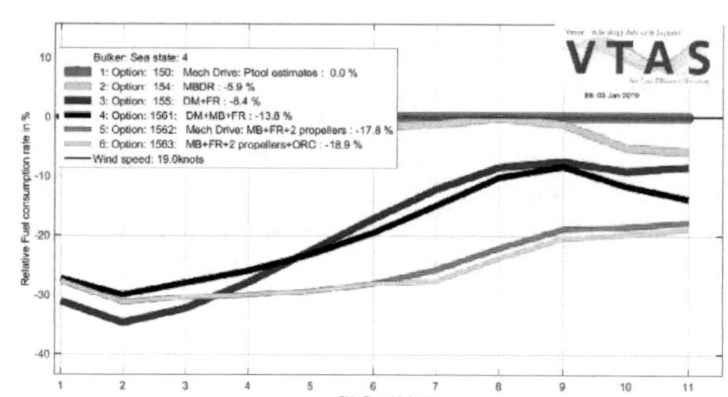

Figure 8. Fuel Consumption Comparison

Figure 8 shows the relative fuel consumption of the set of design options compared to the baseline design. Clearly the options that include adoption of a twin shaft solution show the single greatest impact, illustrated in pink and cyan.

6. CONCLUSIONS

This study shows how the collective application of a set of EST can lead to fuel savings of up to 20% at the service speed of a 61,000 tonne dwt bulker. At lower speeds the benefits increase due to the advantageous use of FR in improved apparent wind conditions.

As more operating information on the performance of various ESTs in different operating conditions becomes available, the individual EST models will be refined to more accurately capture these effects and provide an increasingly reliable estimate of their future performance.

7. ACKNOWLEDGEMENTS

This paper represents work carried out by BMT Defence And Security UK Ltd with the permission of the Energy Technologies Institute's Heavy Duty Vehicle (HDV) Marine programme, whose support in the development of this paper is gratefully acknowledged.

It is BMT's intention to claim copyright for this work. The kind permission and resources granted to the authors by BMT are acknowledged with thanks. All findings, ideas, opinions and errors herein are those of the authors and are not necessarily those of BMT Defence and Security UK Ltd.

8. AUTHORS' BIOGRAPHY

David Pearson holds the position of Senior Engineer at BMT Defence and Security Ltd. His primary domain is marine consultancy and auxiliaries design, with a focus on new and innovative Energy Saving Technologies. His previous experience includes modelling wind-based technologies for the Energy Technologies Institute, among a range of other EST projects.

John Buckingham CEng FIMechE, is the Chief Mechanical Engineer of BMT Defence & Security UK Limited, Bath, UK. John is the Chief Technologist for the Energy Technologies Institute funded Vessel Technology Assessment System project.

John leads studies into marine power and propulsion, HVAC and hydraulics within the company and has published several papers on these subjects. He has also published other papers on a range of subjects from biofuels, organic Rankine cycles to rail-gun cooling. Since 1999, John has been involved in a range of studies to assess the utility of energy saving technologies for the commercial and military marine.

9. REFERENCES

1. Holtrop, J & Mennen, GGJ. "An Approximate power prediction method". Netherlands Ship Model Basin (MARIN). Wageningen, The Netherlands. 1982.

2. Pearson & Buckingham "Modelling The Performance Of Energy Saving Technologies On Ships" RINA Full Scale Ship Performance conference, London Dec 2018.

3. van Lameren, W.P.A., van Manen, J.D. and Oosterweld, M.W.C., "The Wageningen B-Screw Series", Transactions SNAME, v. 77, p. 269-317, 1969.

4. Moriguchi. Y, Kato. H, Influence of microbubble diameter and distribution on frictional resistance reduction, Journal of Marine Science and Technology, September 2002, Volume 7, Issue 2, pp 79–85

5. Dindorf, R. "Estimating Potential Energy Savings in Compressed Air Systems", Procedia Engineering 39:204–21, December 2012.

6. Mizokami, et al. "Experimental Study of Air Lubrication Method and Verification of Effects on Actual Hull by Means of Sea Trial" Mitsubishi Heavy Industries Technical Review Vol. 47 No. 3, September 2010.

7. Mäkiharju et al. "On the energy economics of air lubrication drag reduction", International Journal of Naval Architecture and Ocean Engineering, Volume 4, Issue 4, December 2012, Pages 412-422.

8. Kawakita at al. "Application of Simulation Technology to Mitsubishi Air Lubrication System" 2015.

9. Foeth, E J. "Decreasing frictional resistance by air lubrication". 2009.

10. Lay, K.A., Yakushiji, R., Mäkiharju, S., Perlin, M. and Ceccio, S.L., "Partial cavity drag reduction at high Reynolds numbers", Journal of Ship Research, 54(2), pp. 109-119, 2010.

11. Kodama at al. "Microbubbles: Drag Reduction Mechanism and Applicability to Ships" 2002.

12. Kodama, et al., "Practical application of microbubbles to ships - Large scale model experiments and new full scale experiment" 2004.

13. Pearson. D, 'The Use of Flettner Rotors in Efficient Ship Design', RINA, Influence of EEDI on Ship Design, 24-25 September, London, UK

14. Hydrocurrent Organic Rankine Cycle Module 125EJW -Compact and High-performance Waste Heat Recovery System Utilizing Low Temperature Heat Source- Mitsubishi Heavy Industries Technical Review Vol. 52 No. 4 (December 2015).

CAN BIOFUELS HELP PROVIDE CLEAN PROPULSION FOR SHIPPING, NOW AND IN THE FUTURE?

C Lewis and **A Bauen**, E4tech (UK) Ltd., UK

SUMMARY

E4tech conducted an important study (in 2018) for the Dutch Biofuels Platform and the Port of Rotterdam, as part of the development of a roadmap for decarbonising shipping in the Netherlands. In particular, biofuels can provide large reductions in GHG and non-GHG emissions, with a range of solutions for decarbonisation in the short and longer term. They offer greater decarbonisation potential compared to fossil alternatives aimed at tackling air quality emissions (e.g. LNG). The biofuels analysed were hydro-treated vegetable oil (HVO – including from waste oils and fats), Fatty acid methyl ester (FAME), straight vegetable oil (SVO), ethanol (both conventional and advanced production processes), bio-methanol, bio-LNG, Fischer-Tropsch diesel (FT-diesel) and Upgraded Pyrolysis Oil (UPO).

The study determined the most attractive biofuel options based on criteria such as GHG reduction potential, readiness of production, cost and compatibility with the current vessel fleet in each shipping sub-sector: deep-sea, short-sea and inland shipping. It also provided recommendation on coordinated interventions from the range of actors in shipping industry that can support the uptake of biofuels, overcoming a range of technical, economic and operational barriers to their use.

NOMENCLATURE

CNG	Compressed Natural Gas
EU	European Union
FAME	Fatty acid methyl ester
FT	Fischer-Tropsch
GHG	Greenhouse Gases
HVO	Hydrotreated Vegetable Oil
IMO	International Maritime Organisation
LNG	Liquefied Natural Gas
LPG	Liquid Petroleum Gas
NOx	Nitrous oxides
PM	Particulate matter
SOx	Sulphur Oxides
SVO	Straight Vegetable Oil
UPO	Upgraded Pyrolysis Oil

1. INTRODUCTION

The reduction of greenhouse gas (GHG) emissions to mitigate climate change has become an important global effort, having been given structure, urgency and legitimacy by the Paris Agreement in 2015. The shipping sector is increasingly under pressure at national and international level to reduce greenhouse gas emissions, as well as emissions affecting air quality. A range of alternative fuels exist that can tackle these issues. This paper looks specifically at the use of biofuels.

2. EMISSION REDUCTION DRIVERS

Both inland and sea-faring shipping are currently subject to drivers of change due to non-GHG emissions. The International Maritime Organisation (IMO) and European Union (EU) have applied stringent limits on sulphur oxides (SOx), nitrous oxides (NOx) and particulate matter (PM) emissions [1]. These constraints are greater for inland shipping through the EU regulations [2]. Therefore, any decarbonisation options need to also comply, and if possible advance, non-GHG emission reductions. In April 2018, the IMO adopted a target to reduce GHG emissions from the sector by 50% by 2050 on 2008 levels, with a full strategy implementing measures to reach this goal expected in 2023 [3].

3. OPTIONS FOR REDUCING EMISSIONS

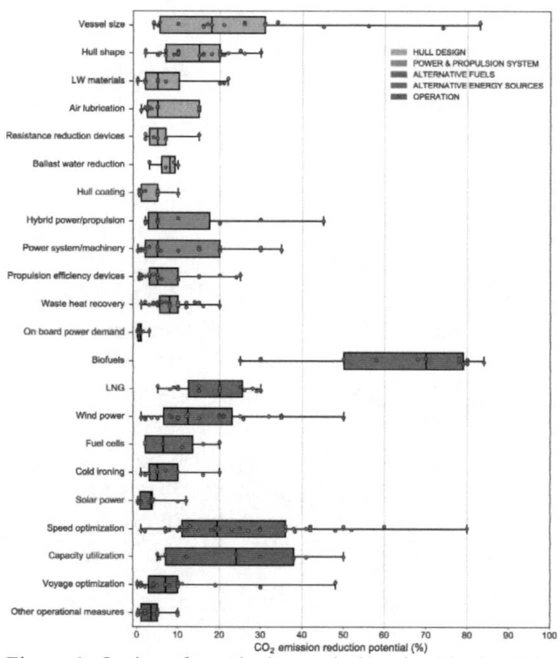

Figure 1: Options for reducing emissions in shipping [4]

GHG emissions in shipping can be reduced in different ways: by changes in operations like slow-steaming; increases in efficiency through stream-lining or energy recovery systems; or using alternative energy carriers for propulsion and operation of the vessels. However, this study focuses on the use of alternative energy carriers and fuels, including fossil fuel and low-carbon fuel options.

Whilst there are some advantages to fossil-based alternatives to fuel oil and middle distillates, such as LNG, CNG, LPG and fossil-based methanol, the GHG reduction from these options is limited to approximately 20% GHG savings [4]. The larger GHG emission savings can only be achieved by using low-carbon energy carriers [5]. The most feasible of these options include the use of electricity and hydrogen (whose emission reductions depend only on the source of production as there are no tailpipe emissions); nuclear (which can achieve very high GHG emissions reduction on a lifecycle basis, but has substantial barriers to implementation due to safety and geopolitical factors); and biofuels [6].

Biofuels can provide large reductions in GHG and non-GHG emissions, offering a range of solutions for decarbonisation in the short and longer term.
The biofuels analysed were hydro-treated vegetable oil (HVO – including from waste oils and fats), FAME, straight vegetable oil (SVO), ethanol (both conventional and advanced production processes), bio-methanol, bio-LNG, Fischer-Tropsch diesel (FT-diesel) and Upgraded Pyrolysis Oil (UPO). These biofuels were qualitatively analysed on their GHG reduction potential, readiness of production, cost and compatibility with the current vessel fleet in each shipping sector. As a reference:

- Deep-sea shipping was assumed to use HFO fuel with scrubbers to comply with SOx regulations tightening to 0.5% sulphur limit in 2020 [1].
- Short-sea shipping assumed an archetype of MGO to comply with the 0.1% sulphur limit [1].
- Inland shipping assumed the use of EN590 diesel due to sulphur requirements of less than 0.001% [8].

3.1 SHORT-SEA AND INLAND SHIPPING

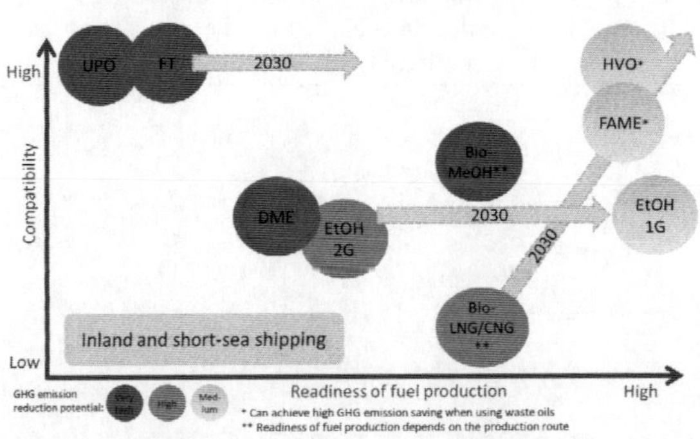

Figure 2: Readiness of fuel production and compability with current inland and short-sea shipping fleet

Fuels for use in short-sea and inland shipping sectors are summarised in Figure 2. HVO, FAME, FT-Diesel and UPO can be described as 'drop-in' biofuels, as they are liquid fuels that are functionally equivalent to their petroleum counterparts in marine engine combustion and are also compatible with current storage and refuelling infrastructure. SVO from vegetable crops was not a viable option for use in inland and short-sea shipping due to the engine conversion requirements and low overall attractiveness (low GHG reduction potential and issues with sustainability) [7].

HVO is currently available at commercial scale, allows very high GHG reductions when using waste oils and fats, making it the most attractive short-term option to decarbonise the shipping sector. The use of HVO from vegetable crop oils raises sustainability concerns, and their use is being restricted by EU policy. However, the 'drop-in' nature of HVO fuel in road and aviation also leads to strong competition from those sectors, and the limited availability of waste oils and fats may limit their use [6].

FAME in contrast to HVO is currently produced in larger volumes, but suffers from the same feedstock constraints, has a blending limit of 7% in inland and short-sea vessels if EN590 diesel standards for inland shipping need to be met, and does not reduce NOx emissions. However, higher blend levels of FAME are technically possible, although implications on the engines and non-GHG emissions need to be carefully considered [6].

FT-Diesel and UPO have a lower commercial readiness, limiting their current availability and in 2030, and implying higher costs. While UPO is at an earlier stage of development, its cost could be relatively attractive, justifying further development of this technology [6].

Conventional ethanol is widely available today but is incompatible with current marine engines, so is not a good short term option. However, the fuel supply and storage can be adapted in current engines to use ethanol or methanol. This could lead to ethanol compatible vessels being operational by 2030. More substantial GHG savings can be achieved through advanced ethanol production, which should be commercially available by 2030, though again costs will be higher due to the early days of commercialisation of the technology [6].

Bio-methanol is potentially a very attractive option in terms of costs and GHG emissions reductions, and is commercially produced today from biomethane. However, the production of bio-methanol via gasification of solid biomass is only at early commercial stage, and the production of methanol from renewable electricity is only at a large demonstration stage. Bio-methanol would be more attractive to the inland and short-sea shipping sectors as the 50% lower energy density of methanol (compared to incumbent fuels) limits the vessel's range, which is a major draw-back for deep-sea shipping. DME, which can be produced from or instead of methanol, requires separate storage and bunkering infrastructure, as it is in a gaseous state in ambient conditions and requires a constant 5bar pressure to remain liquid. Even though DME is more compatible with marine diesel engines than methanol, it still requires adaptation of the engine and fuelling system, and is hence deemed less attractive due to more complicated storage and bunkering [6].

Bio-LNG is commercially available today from biomethane, can achieve high GHG emission reductions, and could be used, either directly or via a mass-balancing system (applied to bio-LNG or biomethane) in existing LNG vessels. When produced from bioSNG, it is at a lower readiness level, due to the ongoing development of bioSNG technologies. There are currently a low number of LNG compatible vessels, and the costs of converting vessels and infrastructure is more expensive than for methanol. If bio-LNG is to be attractive in 2030, a significant increase in the uptake of LNG vessels is needed. Bio-CNG is considered as a potential option in short-sea and inland shipping, but more research is needed on the implications of CNG use on the current vessel fleets [6].

3.2 DEEP SEA SHIPPING

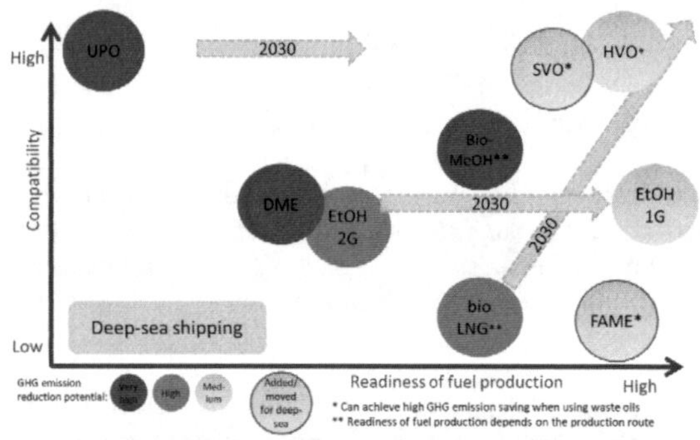

Figure 2: Readiness of fuel production and compability with current deep-sea shipping fleet

The outlook only slightly changes for deep-sea shipping. SVO is compatible with the HFO engines used in deep-sea shipping [7] and is included as a viable drop-in option. Additionally, FAME is only allowed in trace amounts based on the ISO 8217 standards applying to HFO in deep-sea shipping [9].

The price differential between biofuels and HFO used in deep-sea shipping is greater than that of the price differential between biofuels and the fuels used in inland and short-sea shipping. Therefore, biofuels are much less economically viable in the deep-sea sector [6].

4. BARRIERS AND INTERVENTIONS

There are technical, economic and operational barriers to the use of biofuels in shipping, however, coordinated interventions from the range of actors shaping the shipping industry can support the uptake of biofuels.

The use of biofuels in shipping is very limited at the moment [10], but there is the potential for growth via the range of sustainable biofuels described in this study. This will require a set of interventions from different stakeholders, and coordination between them for greatest effect, to address technical, economic and operational barriers. As the short-sea and deep-sea shipping sectors are mainly governed by IMO regulations (and will be influenced by the recent GHG reduction strategy), countries' efforts may be best focused on inland and domestic shipping where it can have greatest impact through policy and interventions on a national level. However, stakeholders can also play an important role in facilitating change in the short-sea and deep-sea sectors – in particular, by working with ports, who play a key role at the interface between the deep-sea, short-sea and inland shipping sectors.

Table 1: Main barriers to biofuels in shipping and intervention actions for important stakeholders [6].

Barriers	Actions	Actors
Limited experience with and availability of alternatively fuelled vessels and related refuelling infrastructure	Evaluation of current experience and dissemination of best practice	Government in collaboration with key concerned parties
	Demonstration projects	Government or international / Dutch ports coalition
	Infrastructure grants / finance	Government initiatives to leverage private sector funding
	Port levy to (co-)fund infrastructure	Ports
	Fuels procurement commitments	Shipping industry
	Incentives for transition to alternatively fuelled vessel fleet	Government, ports

Price differential with fossil fuels	Advanced biofuels sub-target for inland shipping. Multiplier for use of advanced biofuels in shipping	Government
	Carbon pricing. Linking tonnage taxes to decarbonisation	Government
	Public procurement of biofuels	Government, public sector
	Reduction in port fees and other incentives at ports	Ports
	Private sector procurement of biofuelled shipping services	Industry (goods)
Lack of certification regulation	IGF codes for ethanol and methanol	IMO, classification bodies
Lack of policy alignment	Initiatives to align policies at European level and between different countries	Government, IMO, EU
	Mass balancing system for biomethane or bio-LNG	Government, EU
Lack of information concerning biofuels use	Information sharing, for example via establishment of a shipping biofuels forum	Relevant industry, associations and government departments

7. CONCLUSIONS

Biofuels provide options to decarbonise the shipping sector in both the short term and long term, with greater emissions reductions being achieved by the fuel production routes that are still developing to be commercial technologies. Biofuels, in many cases, offer compatible options for vessels where large changes to engine and on-board storage infrastructure.

There are many barriers that remain for the implementation of these fuels, not limited to the commonly discussed barriers of fuel cost and availability. Stakeholders that can intervene to overcome these barriers have been identified.

8. ACKNOWLEDGEMENTS

This conference paper has been adapted from a full report produced for the Platform Duurzame Biobrandstoff (Dutch Platform for Sustainable Biofuels) and our thanks goes to PDB for commissioning this research. The full report can be found at:

https://platformduurzamebiobrandstoffen.nl/infotheek-item/masterplan-for-co2-reduction-in-dutch-shipping-sector-biofuels-for-shipping/

9. REFERENCES

References should be numbered in the text where they occur, with the number in square brackets [1] and a complete list given at the end of the paper.

Cross Referencing, footnotes, endnotes and 'Smart Tags' must NOT be used for figures and/or references.

1. International Maritime Organisation, 'Prevention of Air Pollution from Ships', http://www.imo.org/en/OurWork/Environment/PollutionPrevention/AirPollution/Pages/Air-Pollution.aspx, 2019.
2. European Union, 'Regulation (EU) 2016/1628 of the European Parliament and of the Council', Official Journal of the European Union, 2016
3. International Maritime Organisation, 'Adoption of the Initial IMO strategy on reduction of GHG emissions from ships and existing IMO activity related to reducing GHG emissions in the shipping sector', https://unfccc.int/sites/default/files/resource/250_IMO%20submission_Talanoa%20Dialogue_April%202018.pdf, 2018.
4. Bouman, E. A., Lindstad, E., Riall, A.I, and Strømmana, A.H., 'State-of-the-art technologies, measures, and potential for reducing GHG emissions from shipping – A review', Transportation Research Part D: Transport and Environment. Elsevier, Volume 52, Part A, May 2017, Pages 408-421, doi: 10.1016/j.trd.2017.03.022, 2017.
5. DNV GL, 'Assessment of Selected Alternative Fuels And Technologies', https://www.dnvgl.com/maritime/publications/alternative-fuel-assessment-download.html, 2018.
6. E4tech, 'Master plan for CO2 reduction in the Dutch shipping sector - Biofuels for shipping, https://platformduurzamebiobrandstoffen.nl/infotheek-item/masterplan-for-co2-reduction-in-dutch-shipping-sector-biofuels-for-shipping/, 2018.
7. IEA Bioenergy, 'Biofuels for the marine shipping sector', https://www.ieabioenergy.com/publications/biofuels-for-the-marine-shipping-sector/, 2017.

8. European Union, 'Directive 2009/30/EC of The European Parliament and of the Council' http://eur-lex.europa.eu/legal-content/EN/TXT/PDF/?uri=CELEX:32009L0030&from=EN, 2009.
9. International Organization for Standardization, 'ISO 8217:2017 Petroleum products -- Fuels (class F) -- Specifications of marine fuels', https://www.iso.org/standard/64247.html, 2017.
10. International Energy Agency, 'Tracking Progress: Transport biofuels'. Available at: https://www.iea.org/ctp/tracking2017/transportbiofuels/, 2017.

9. AUTHORS BIOGRAPHY

Chester Lewis holds the current position of Consultant at E4tech. At E4tech Chester works on low-carbon fuels for application in the maritime industry. His previous experience includes research in the fields of energy systems, fuel cells and hydrogen and low-carbon transport projects.

Ausilio Bauen holds the current position of Director at E4tech. His previous experience includes extensive research and consulting experience on technical, economic, environmental and policy issues relating to decentralised generation and alternative fuel production and infrastructure. His focus is on biomass energy, fuel cells and related fuels for stationary and transport applications, and decentralised generation integration into energy systems. Ausilio has worked with industry, government, non-government and international organisations.

EXAMINING METHANOL AS AN ALTERNATIVE MARINE FUEL FOR INDONESIAN DOMESTIC SHIPS

E M Priyanto, Biro Klasifikasi Indonesia, Indonesia
A I Ölçer, D Dalaklis, and **F Ballini,** World Maritime University, Sweden

SUMMARY

The analysis in hand aims to provide insight and to explore the future usage of methanol as an alternative marine fuel for domestic ships in Indonesia. An overview of potential application, analysis of resources availability, and stakeholder readiness/overall views on the topic are provided; related challenges are also identified and further examined.

The potential performance of methanol as a fuel is discussed and evaluated via two different perspectives (the ship-owner perspective and the government one), through case studies of two passenger ships owned by the shipping company Pelayaran Indonesia (PELNI): MV. Labobar and MV. Gunung Dempo. As ship-owners tend to look very closely at the economic aspects, a feasibility study is performed by developing a combinatorial scenario approach based on the combination of economic measures of merit (NPV and payback period) along with a technical scenario (main-pilot fuel set up); a few of the variables included in the calculation are: ship age, ship productivity, and macro-economy conditions. Regarding the government perspective, the issues of environmental protection and policy compliance are evaluated by examining six emission types (NOx, SOx, CO_2, CH_4, N_2O, and PM). Additionally, since there is a trade-off situation in government subsidies between the government and ship-owner interests, an optimization and sensitivity analysis is performed by utilizing a combinatorial scenario model to determine optimum methanol price and external variables influencing the decision to support further use of methanol in the Indonesian market.

An important finding was that Indonesia has certain advantages/drivers to introduce methanol as marine fuel. However, methanol competitiveness is mainly dependent on ship productivity and the price differences between methanol and marine diesel oil (MDO). Additionally, policy analysis (through an optimization approach) could be one of the government options in order to determine the optimum condition in establishing methanol as marine fuel. Finally, short, medium, and long term recommendations are also provided as the basis for future consideration.

1. INTRODUCTION

Indonesia has recently experienced a significant increase in energy consumption: from 508.88 mBOE in 2000 to 830.42 mBOE in 2015; this translates into an annual increase of 3.99% for the specific time-frame [14][32]. Meanwhile, this country is still very heavily dependent on fossil fuels and non-renewable energy supply, considering that approximately 46% of the necessary resources are supplied by fuel oil [14]. In any case, increasing fossil fuel consumption, declining refinery production, and no new discovery of oil reserves resulted into a change of the import ratio of oil products and crude oil, which was increased from 33% in 2006 to 44% in 2015 [26]. By using the assumption of the reverse-production ratio of fossil fuel in 2014, then, oil will be depleted in 12 years and coal in 70 years; also, natural gas will run out in 37 years [32]. This condition could lead Indonesia into becoming a net-energy importer and highly vulnerable in terms of national energy security.

Meanwhile, the transportation sector is the largest fuel oil consumer in Indonesia, holding a share of around 329.039 mBOE, or 86% of the total fuel oil consumption [14]. Furthermore, the wider shipping industry in the country is expected to grow significantly in the future, especially when factoring in the new policy from the Indonesian government which concentrates on port infrastructure development and domestic shipbuilding to support the vision of Indonesia's global maritime nexus.

Nevertheless, an increase of fuel oil consumption for serving maritime transport needs will also contribute into an increase of green-house gasses (GHG) emissions and further downgrade the quality of air. At the same time, high demand, volatility of oil price, and stringent environmental regulation could also create disturbances in the normal conduct of shipping operations. Since maritime transport is the backbone of economic development in Indonesia, it is necessary for the Government to address and encourage the utilization of sustainable solutions for energy transition and incorporated initiatives in its future energy policies.

For example, one solution could be identified in the introduction of methanol as a marine fuel. Sustainability of feedstock provides methanol with an advantage over other alternative fuels, since it can be used both as a transitional marine fuel and future sustainable option. Unfortunately, in the wider literature and projections [32][26], Methanol has not yet been acknowledged as a promising future alternative fuel in Indonesia.

2. EXAMINING POSSIBLE FEEDSTOCK OF METHANOL AND METHODS OF PRODUCTION

Methanol is commonly referred to as wood alcohol (or, methyl alcohol), with the chemical formula CH_3OH. For reasons of convenience (terminology used in the market), it is quite often abbreviated as MeOH [22]. Methanol can be produced from fossil based resources (non-renewable), or via materials/resources that are considered as renewable and sustainable, such as wood, biomass, sewage, and even from CO_2 [8].

To cut a long way short, there are three basic steps commonly used to produce methanol, namely synthetic gas (syngas) production, syngas to methanol conversion, and distillation or purification of effluent. The sources of synthetic gas can be natural gas, coal, biomass, crude oil, or other carbon based sources. Despite this, the industry still prefers natural gas or methane as the feedstock, since the production cost, energy consumption, and impurities are considerably lower than all the other available options [6]. Indonesia has many potential feedstock-possibilities, either from fossil or renewable resources to facilitate methanol production. The fact that methanol can be produced from various alternatives is making it initially suitable as a transition option and subsequently as a "future sustainable" alternative fuel to effectively serve the country's needs.

2.1 FOSSIL RESOURCES

2.1 (a) Coal

Coal can be used for steam power generation; it can also be treated as a potential methanol feedstock. Methanol can be produced from coal through gasification to produce synthesis gas and then followed by methanol synthesis and purification. Moreover, this type of production will consume 1,42-1,59 ton of coal per ton of methanol [35]. In addition, Indonesia has abundant coal resources, considering that this country is one of the biggest coal producers in the world [16]. In 2015, coal production was 126,61 billion tons, with an additional 32,26 billion tons kept as reserves [15].

2.1 (b) Natural Gas

Methanol production using natural gas in Indonesia has commercially started in 2000 by Kaltim Methanol Industry (KMI), with a production capacity of 600.000 ton per year. In the production of methanol, KMI has been using steam reformer and low-pressure synthesis methanol technology (Figure 1). Approximately 750-1300 m3 of natural gas were consumed to produce one ton of methanol and this outcome is highly depending on the maturity of the technology used [28]. Furthermore, in order to optimize efficiency, the methane slip during the steam reforming processing is processed by using the partial oxidation method [18].

According to Indonesia's Ministry of Energy and Mineral Resources (ESDM) data, the country's total natural gas reserves in 2015 amounted to 150,39 TSCF. The largest reserves are in Natuna (with 50,48 TSCF) and then in West Papua (with a total of 23,90 TSCF). Moreover, Indonesia has other potential natural gas resources, such as shale gas and coal-based methane [26]. Even though this country is rather rich in natural gas resources, the domestic absorption of natural gas is still relatively low [32]. Utilization of methanol as a marine fuel could certainly create an impact in case an increasing domestic gas market is established at the same time.

2.2 RENEWABLE RESOURCES

2.2 (a) Industrial Waste

Indonesia is one of the biggest crude palm oil (CPO) producers in the world [21][34]. The CPO Industry produces a lot of solid waste, such as empty fruit bunches, fibres, and shells, as well as pome [32]. According to Goenadi et al. (as cited in [32]), every ton of palm oil fruit will produce 180 kg of fibres and shell, and 600-700 pome. Decomposition of pome in anaerobic condition may produce biogas containing 50-75% of methane that potentially can be used as a methanol feedstock [34]. At the same time, the sugar industry could potentially be used as a methanol feedstock because of available by-products such as molasses, bagasse, and leaves of cane tops [32][4]. Another industry of interest is the pulp and paper one, which is creating by-products of non-condensate gas that can be used as bio-methanol feedstock [32].

2.2 (a) Municipal Waste

With more than 250 million people as population, Indonesia has a problem with the issue of municipality waste. One of the solutions is to transform this waste into a source of energy. Through the process of sanitary landfill and anaerobic digestion, municipality waste can produce methane as methanol feedstock [32]. However, the concept of utilizing biomass to convert into energy is not popular in Indonesia and the technology that is needed must be imported from outside; also, the culture of citizens sorting out their rubbish to facilitate further process is relatively low.

3. CHALLENGES IN IMPLEMENTATION

3.1 REGULATION CONSIDERING METHANOL AS A MARINE FUEL

For the time being, the existing requirements and standards of marine fuel (regulated by ESDM) only cover fossil fuel and biodiesel. Moreover, Directorate General of Sea Transportation (DGST) has not developed the related requirement for ships using low-flashpoint fuel yet. On the positive side, Biro Klasifikasi Indonesia

(BKI) (as the only classification society that has received full authority from the Government) has already established the necessary regulations for methane-fuelled vessels, based on the IGF code. However, these regulations need to be amended and also to include provisions considering other low-flashpoint fuel, including methanol. According to the IGF Code, "*In the meantime, for other low-flashpoint fuels, compliance with the functional requirements of this Code must be demonstrated through alternative design*". This means that if the Government does not have any prescriptive rules for other low-flashpoint fuel applications (including methanol) then the ship design has to be approved as an alternative design, through risk assessment. Although BKI has developed the guidance for risk evaluation for an alternative arrangement, the Indonesian Administration does not possess such regulations. Cooperation between both institutions is highly needed for the success of the implementation of methanol as marine fuel in Indonesia.

There is also an absence of regulations regarding authorization, standardization, and certification of ship bunker suppliers in Indonesia. Currently, BKI and DGST are working together to establish a scheme to maintain the quality of marine fuels and to promote the availability of domestic fuels based on MARPOL Annex VI Regulation 18. Since this standardization is still an on-going process, additional types of alternatives fuels, such as methanol, can also be considered.

3.2 TRAINING AND COMPETENCY

Despite its potential advantages as a new alternative fuel technology, the application of methanol as a marine fuel may confer risks to a person on-board the vessel. There is also the possibility of mishandling by the crew due to unfamiliarity or lack of training. It is important to keep in mind that there might be a reluctance to use new fuel systems by "traditional" seafarers. There is no maritime institute in Indonesia with the necessary facilities providing appropriate training and certification of proficiency based on the IGF Code (and the STCW).

To overcome the above challenges, the Maritime Administration needs to establish a compact training module, which consists of a theoretical and a practical program based on the STCW Convention Part A-V/3 of regarding "Mandatory minimum requirement for the training and qualification of master, officers, rating, and other personnel on ships subject to the IGF Code". Moreover, the concerned shipping company must ensure the familiarization of the crew on-board by having appropriate exercises/drills according to the ISM Code, especially on the use of personnel protection equipment.

3.3 COORDINATION AMONG STAKEHOLDER

In 2014, the Indonesian government established the "Coordinating Ministry of Maritime Affairs" that coordinates and synergizes 4 different other Ministries (Ministry of Transport, Ministry of Fisheries, ESDM, and Ministry of Tourism). However, duties are merely on coordination, whilst there is no authority-obligation on providing a strategic maritime energy roadmap and policy making that is associated with all the respective ministries and institutions [12].

Indeed, establishing an energy policy and introducing methanol as an alternative energy source in the country's maritime transport industry requires coordination among the stakeholders and preferable should not only be handled by the Ministry of Transportation. According to the regulation of the Ministry of Energy and Mineral Resources No.45/2005, regarding the standards and quality and supervision of various fuels that are marketed domestically, the authority to manage and standardize the quality, technical provision, and availability of marine fuels lies on the Ministry of Energy and Mineral Resources. However, the Ministry of Transportation requires the data of fuel availability and quality to comply with the respective MARPOL Annex VI requirements. In addition, the strategy to introduce methanol as fuel into maritime industry should also include the ship conversion activity (see Figure 1).

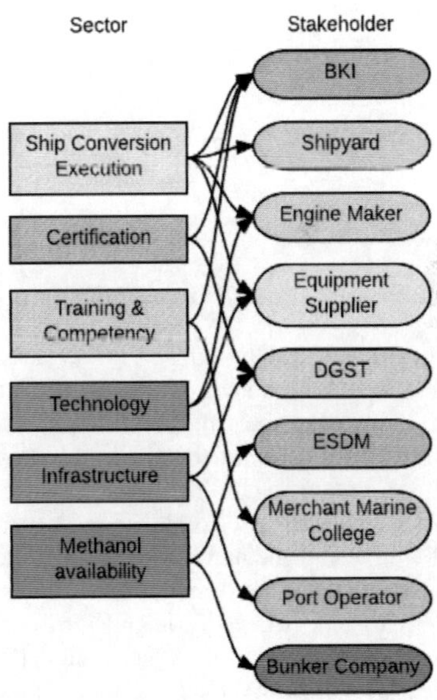

Figure 1: Stakeholders in the conversion activity of methanol-fuelled vessels

3.4 SUPPLY, DEMAND, AND LOGISTIC

Total supply of methanol in Indonesia in 2014 was around 450.000 tons, and it was mainly produced from KMI [13]. Considering that KMI has the capability to produce methanol up to 600.000 tons, there is an opportunity to increase its production (if off-course the market can absorb this). According to the market projection from the Ministry of Industry, in 2020 the total methanol demand in Indonesia will be 2,4 mta [13].

Currently, 80% of methanol demand in this country is derived from the formaldehyde industry [18]. Developing methanol as a marine fuel in Indonesia can improve absorption and even create a totally new energy market in relation to the maritime industry. Even though methanol can be an energy resource, there is no establishment of such a market (Figure 2). Nevertheless, establishing this new market requires enormous efforts and stiff cooperation among all the stakeholders, which are located within different sectors.

(HFO, MDO, MGO) in the infrastructure of bunkering, distribution, and storage. However, since methanol is a low-flashpoint fuel, then there are certain minor modifications necessary, in case the existing marine fuels infrastructure is chosen for supporting the specific fuel option. In addition, at the beginning of methanol implementation, there is no real need to have the new infrastructure in place, since the bunkering process can be performed by using the "truck to ship" method [20].

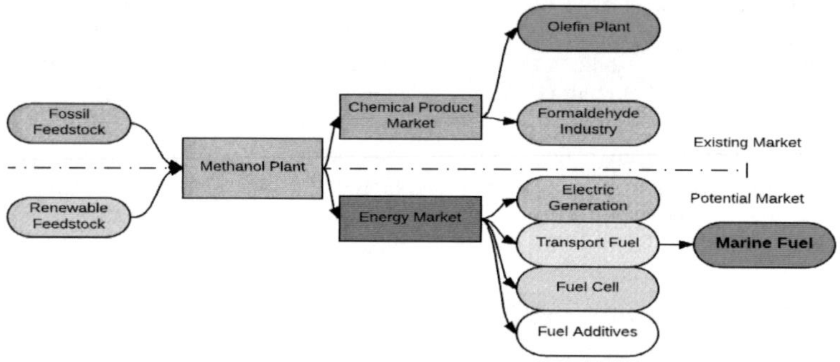

Figure 2: Existing and Potential Market of Methanol in Indonesia (Authors, 2018).

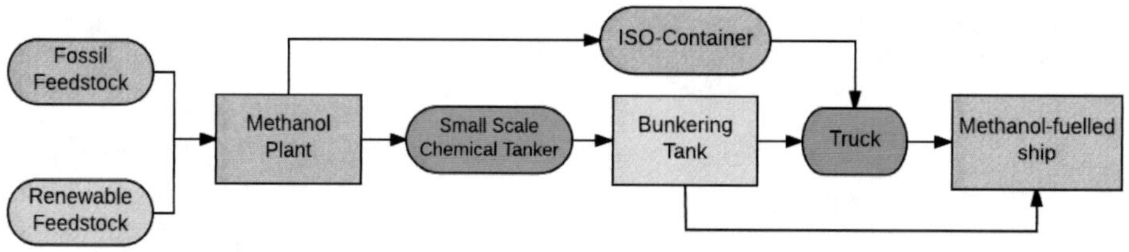

Figure 3: Simple Supply-Chain of methanol as marine fuel (Authors, 2018)

On the other hand, as Indonesia is an archipelagic country, the logistic channels need to be established according to the market assessment in the targeted island. There are three option for methanol distribution: small-scale chemical tanker, ISO-tank container, or by truck (Figure 3). Small-scale vessel or ISO-tank container can be used for delivering methanol in long distances from the producer or to large consumer islands; trucks can be used for short distances/near-by islands. In addition, to improve the future market, it is necessary to provide sufficient fleet of methanol tankers, to meet the demand.

3.4 INFRASTRUCTURE

To maintain the supply-chain and the availability of methanol as a marine fuel, infrastructure in designated ports will be needed [1]. Currently, the existing methanol infrastructure is used for suppling the chemical industry: the main port for loading belongs to KMI with a capacity of 30.000DWT and most unloading is taking place in the PT. Siam Maspion Terminal.

Since methanol has liquid properties in atmospheric pressure, there are similarities with existing marine fuels

4. ENVIRONTMENTAL AND TECHNO-ECONOMIC ANALYSIS

Introducing methanol as a marine fuel in the Indonesian market requires investments and clarity in the national strategic energy and transport roadmaps, for instance a government subsidy system. If the provisions to support this application have not been established in advance, implementation can be significantly hampered [9]. In order to understand to what extent methanol can be introduced to the market, a techno-economic and policy-making analysis is performed via the case study of two passenger ships owned by Pelayaran Indonesia (PELNI).

The discussion has been divided into two perspectives: the ship-owner perspective and the government one. Typically, ship-owners look at the economic aspects and benefits, such as Net Present Value (NPV) and the payback period. On the other hand, the government rather considers the optimum support to the market, including subsidies, as well as compliance with established regulations.

4.1 SHIP-OWNER PERSPECTIVE

From the ship-owner perspective, retrofitting an existing ship is preferred than building a new one, since they are emphasizing on economic considerations [2]. Moreover, market conditions such as over-supply, volatility of oil price, and stringent regulations can make a ship-owner more cautious to invest in new ships. Therefore, in order to understand the effectiveness of methanol technology investment on main engine and which ship is possible to be retrofitted, then a case study of technology investment behavior towards the ship revenue has been performed on two PELNI passenger ships: MV. Labobar and MV. Gunung Dempo.

4.1 (a) Case study of PELNI passenger ship

MV. Labobar is a T-3000 type that is capable of loading up to 3000 passengers, while MV. Gunung Dempo is a T-2000 type (see Table 1). Both of them are 2-in-1 ships which are capable of loading both passengers and cargo. PELNI also employs the T-1000 type. However, the last one is excluded from the case study since the estimated conversion cost is based on a passenger ship with 10-25 MW main engine.

Table 1: Ship particular of MV. Labobar and MV. Gunung Dempo [27][5]

No	Parameter	Labobar	Gunung Dempo
1	Type	T-3000 (2-in-1)	T-2000 (2-in-1)
2	Passenger	up to 3000	up to 2000
3	Container cargo	up to 28 TEUS	up to 98 TEUS
4	DWT	4238	3482
5	GT	15136	14017
6	LOA	146.3	147
7	Year of Built	2004	2008

There are three major cost variables and one benefit variable in this economic study, namely "capital cost", "opportunity cost", and "operational cost" as cost variables and "earning" as benefit.

- Capital costs are the investment or fixed costs incurred in the engine conversion activity including the shipyard cost, procurement of equipment, and retrofitting cost. In this study, all parameters in capital costs were incorporated into a single cost as the function of cost/kW.

- Opportunity cost is the loss of revenue due to the retrofitting activity. A retrofitting activity conducted in a shipyard will result in loss of revenue associated with a certain trip. However, the opportunity cost is reduced due to idle fuel cost (which is unused during the retrofitting period). Moreover, in this study this cost is incorporated to the Capital expenditure (Capex), which is represented by the following formula: Capex = Capital cost + Opportunity cost – Total idle fuel cost
- Operational cost is the cost that arises during main engine operation, including operation-maintenance costs and fuel costs. The operational cost increases year by year because it goes hand in hand with the inflation. For main engine fuel cost, it is calculated as follows:

$$\text{Fuel Cost}_{methanol} = [\text{Price}_{methanol}] \times [\%\text{ of methanol}] \times [\text{Fuel Consumption}_{MDO}] \times \frac{\text{LHVMDO}}{\text{LHVMethanol}}$$

$$\text{Fuel Cost}_{MDO} = [\text{Price}_{MDO}] \times [\%\text{ of MDO}] \times [\text{Fuel Consumption}_{MDO}]$$

$$\text{Total Fuel Cost} = \text{Fuel Cost}_{MDO} + \text{Fuel Cost}_{Methanol}$$

- The benefit is the saving for the ship-owner due to operating with methanol. It is represented by the difference in fuel cost that included in the earning before taxation and depreciation (EBTD) as the following formula: EBTD = Revenue – Operational Cost

4.1 (b) Scenario and assumption

In order to identify investment behaviour of methanol, two scenarios have been considered:

- The composition of methanol as main fuel and MDO as pilot fuel. Referring to previous research, Srivastava [32] used M-85 (85% methanol - 15% distillate fuel) for the respective scenario calculation. According to Laakso [19], the use of oil fuel as pilot fuel was lower compared to the methanol as main fuel, but the difference might be related to methanol properties used in the specific ship. Since the use of methanol as marine fuel is relatively novel, the situation might improve in the future. In this study, the scenario of the composition of methanol as main fuel and MDO as pilot fuel will be M-80, M-85, M-90, and M-95. Moreover, as a comparison, the scenario of 100% MDO will also be calculated.

- The percentage of methanol price compared to MDO. It is difficult to determine the pattern of fuel price since the issue of pricing is rather volatile and very unpredictable. However, price history can be used to estimate the future behavior of methanol and MDO price. According to the methanol-MDO price history from 2004-2016 (see Table 2), the highest percentage was 73,02% in 2004, and the lowest was 43,69%. Almost the percentage of methanol over MDO was in 40-60%, hence those percentages range with interval 2% is used as the basis of the scenario. Furthermore, the MDO price in 2016, USD 460,74/tons or Rp 6.136.596,06/tons, is used as the basis of the techno-economic calculation and combinatorial scenario analysis.

Table 2: Percentage of methanol-MDO price history from 2004-2016 [10][11]

Tahun	MDO bunker prices, Singapore ($/tons)	Methanol Price ($/tons)	% Methanol : MDO Price
2004	372.50	272	73.02%
2005	576.00	260	45.14%
2006	645.00	420	65.12%
2007	670.63	300	44.73%
2008	919.50	500	54.38%
2009	566.25	280	49.45%
2010	638.25	325	50.92%
2011	926.00	470	50.76%
2012	972.75	425	43.69%
2013	902.94	470	52.05%
2014	869.06	420	48.33%
2015	498.69	315	63.17%
2016	460.74	275	59.69%

In addition, there are certain assumptions required to perform these calculations:

- The ship maintenance costs remain similar when comparing "before" and "after" the conversion. As methanol is considered as a clean fuel compared to fossil oil fuel, the lifetime of lubricating oil and major spare parts remains equivalent at the same energy efficiency and output as of a diesel engine [31];

- The cost for methanol conversion is taken as 300 EUR/kW as an assumption. According to Stefenson (2014), the cost for methanol conversion was around 300 EUR/kW for Stena Germanica. Moreover, retrofitting costs from diesel into methanol-diesel fuel have been evaluated to be 250-350 EUR/kW for large engines around 10-25 MW [1];

- The average exchange rate used is Rp.13319/USD and Rp.14630/EUR [3];

- The conversion started in the year-end of 2016;

- Depreciation was taken as straight line. This means that the invested methanol technology cost is uniformly reduced through the remaining lifecycle of the ship.

4.1 (c) Combinatorial scenario analysis of NPV calculation

Net Present Value (NPV) represents to what extent a project will increase a company's value. NPV calculated based on the following formula:

$$NPV = \sum_{t=0}^{n} \frac{CFt}{(1+r)^{t}} - \text{Capital cost}$$

Where r is the project's risk-adjusted of capital cost or discount rate, n is remaining economic life, and CFt is the net cash flow at time t that is calculated as EBTD – (Tax x EBT).

NPV is considered as one of the best criteria for investment decisions from a company perspective. When the positive NPV is obtained in a project calculation, it will add value to the company and vice-versa [7].

Next, one of the example NPV calculations of MV. Labobar, with combinatorial scenario of 40% methanol price to MDO and 95-5% composition methanol-MDO, is provided:

$NPV_{(40\%;95\text{-}5)}$ =]$NPV_{year\ 1}$] +...[$NPV_{year\ t}$] – [Capital Cost]

Where,

$CF_{year\ 1}$ = EBTD – 15% EBT
= [Rp26,135,414,969.409] – [15% x Rp21,649,080,001.65]
= Rp22,888,052,969.161

$NPV_{year\ 1}$ = $CF_{year\ 1}$/ (1+ 0.08)1
= Rp22,888,052,969.161/ (1.08)
= Rp21,192,641,638.112

The other NPV is calculated as above until the end of economic life of ship.

$NPV_{(40\%;95\text{-}5)}$ = [Rp21,192,641,638.112]+.....+ [Rp10,997,383,736.250]–[Rp80,754,029,420]
$NPV_{(40\%;95\text{-}5)}$ = Rp199,376,909,066.557

Another NPV calculation with combinatorial scenario is calculated based on the above steps, both for MV. Labobar and MV. Gunung Dempo.

Figure 4 demonstrates the scenario analysis of NPV of MV. Labobar in specific criteria. The investment needed for retrofitting a ship is feasible at any composition of methanol-MDO, when the percentage of the methanol price toward MDO does not exceed 52%. On the other hand, investment in methanol technology in MV. Gunung Dempo is feasible at any scenario given as shown in Figure 5, as all of the NPVs at any scenario (relating to MV. Gunung Dempo) are positive.

From both these figures, the changing behavior of NPV towards the percentage of the methanol price can be seen that the improvement of the payback period is directly proportional to the increment of methanol composition as main fuel up to 46% of the methanol price to MDO. However, from 48% above the trend, the situation will be the other way round.

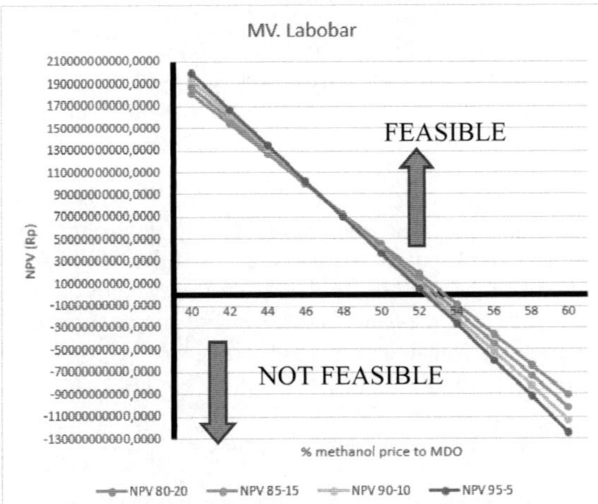

Figure 4: Scenario analysis of NPV-percentage of methanol composition-percentage of methanol price on MV. Labobar case.

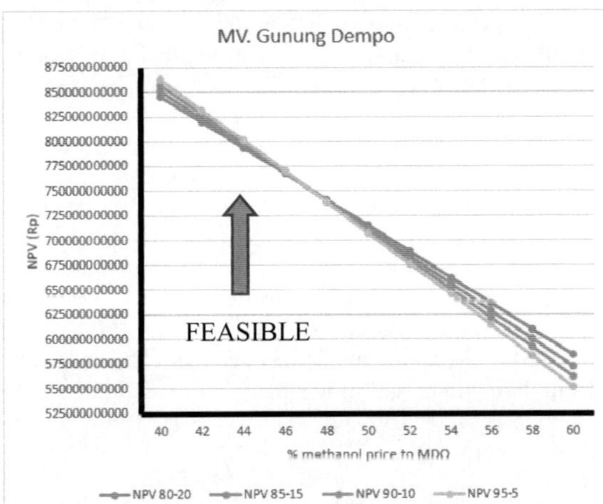

Figure 5: Scenario analysis of NPV-percentage of methanol composition-percentage of methanol price on MV. Gunung Dempo case.

4.1 (d) Combinatorial scenario analysis of payback period calculation

Together with NPV, the payback period is one of the most suitable criteria that has to be considered within an investment analysis. The payback period is the time required when an investment (or, capital cost) is recovered from the operating cash flow and indicated with a positive payback rate. In this study, the payback rate is calculated from discounted cash flow or present value (PV) toward capital cost. The payback period position is in between positive and negative cumulative discounted cash flow [7]. For the remaining period after the last negative payback rate, it is calculated as Present value of the first positive payback rate divided by 12, then multiplied with the number of months, which gives the first positive value when added to the last negative payback rate.

Next, one example of the NPV calculation of MV. Labobar, with a combinatorial scenario of 40% of the methanol price to MDO and 95-5% composition methanol-MDO, is provided:

First positive Payback rate = [year of last negative payback rate (in year 4)] + [PVpositive/12 x 1 month]
= (-Rp706,164,030.81)+(Rp18,145,354,158.192/12 x 1)
= Rp805,948,815.70

So, the payback period for these specific scenarios is 4.1 or 4 year and 1 month. Another determination of the payback period with the combinatorial scenario is calculated based on the above steps, both for MV. Labobar and MV. Gunung Dempo.

More importantly, PELNI does not use a corporate maximum payback time limit in technology investments for ships in decision-making processes [27]. Therefore, Table 3 in this study is employed as an analysis tool in order to determine how feasible methanol conversion is for a shipowner, where the colour also represents the payback time.

Table 31: Colour level of payback period of investment

No	Colour Level	Description
1	Highly Recommended	The payback time is ≤ 1/4 of remaining life cycle of the ship, **and** NPV is ≥ 3/2 of capital cost
2	Recommended	The payback time is 1/4<x≤1/2 of remaining life cycle of the ship, **and** NPV is 1≤x<3/2 of capital cost
3	Not Recommended	The payback time is 1/2<x≤3/4 of remaining life cycle of the ship, **and** NPV is 1/2≤x<1 of capital cost
4	Not Feasible	The payback time is > 3/4 of remaining life cycle of the ship, **and** NPV is < 1/2 of capital cost

Table 4 represents the combinatorial scenario analysis of the payback period of MV. Labobar in the applied scenarios. When examining the associated results, retrofitting of MV. Labobar is highly recommended at any methanol composition when the percentage of the methanol price compared to MDO is 40%. Moreover, it is also advisable to retrofit at instances when the methanol price is up to 46% compared to MDO. However, when looking back to 2016 conditions, in the case that the percentage of methanol over MDO was 59.69% and also by factoring in the revenue condition of MV. Labobar, there is the conclusion that converting the ship into methanol as fuel is not feasible in any scenario from a ship-owner's perspective.

The condition might be changed if the ship-owner can improve the revenue, for instance by improving container cargo capacity (MV. Labobar is a 2-in-1 ship, passenger and container cargo), or getting subsidies from the government for willing to implement green technology (this will be discussed in the government perspective below). In addition, the payback period changed with the percentage of methanol price. Up to 46%, the improvement of the payback period is directly proportional to the increment of methanol composition as main fuel. However, from 48% the trend will be the opposite way.

Table 4: Combinatorial scenario analysis of payback period-percentage of methanol-percentage of methanol price on MV. Labobar case.

Methanol %	Methanol (USD/ton)	80-20	85-15	90-10	95-5
40	184	4.5	4.3	4.2	4.1
42	194	4.11	4.10	4.9	4.8
44	203	5.8	5.7	5.6	5.5
46	212	6.7	6.7	6.6	6.6
48	221	7.11	8	8.1	8.1
50	230	10	10.4	10.7	10.10
52	240	13.8	14.6	15.5	16.7
54	249				
56	258				
58	267				
60	276				

Table 5 represents the combinatorial scenario analysis of the payback period of MV. Gunung Dempo in the stated scenarios. Retrofitting of MV. Gunung Dempo is possible in all applied scenarios. Eventhough MV. Gunung Dempo is smaller than MV. Labobar in terms of size and passenger capacity, MV. Gunung Dempo can gain higher revenues from cargo than MV. Labobar. It can be concluded that, from a ship owner's perspective, the decision of retrofitting a ship to running on methanol also depends on how "productive" the specific ship is.

Table 5: Combinatorial scenario analysis of payback period-percentage of methanol-percentage of methanol price on MV. Gunung Dempo case.

Methanol %	Methanol (USD/ton)	80-20	85-15	90-10	95-5
40	184	1.1	1.1	1.1	1
42	194	1.1	1.1	1.1	1.1
44	203	1.1	1.1	1.1	1.1
46	212	1.2	1.2	1.2	1.2
48	221	1.2	1.2	1.2	1.2
50	230	1.3	1.3	1.3	1.3
52	240	1.3	1.3	1.4	1.4
54	249	1.4	1.4	1.4	1.4
56	258	1.4	1.5	1.5	1.5
58	267	1.5	1.5	1.6	1.6
60	276	1.6	1.6	1.6	1.7

4.2 GOVERNMENT PERSPECTIVE

Ensuring welfare for people and compliance with regulation for all maritime stakeholders is under the government's considerations, while developing business in the maritime sector. In the first attachment of the Presidential Regulation No. 16 of 2017 on Indonesian Maritime Policy, it is stated that the challenge in developing maritime countries is to build inter-regional connectivity and to optimize sea transportation to eliminate social and economic disparities and to facilitate the movement of people, goods, services, and capital. On the other hand, the efforts to increase maritime activities will have negative environmental impacts. Therefore, the government needs to make an effort through robust measures and policy, such as market-based intervention and regulations, to help stakeholders in improving their capability to comply with "green regulations". One indicative form of market-based intervention is by providing subsidies when applying green technology within the maritime sector [33].

In this section, the impact of methanol technology implementation on improvement of environmental protection and policy compliance will be evaluated. Furthermore, an optimization and sensitivity analysis will be conducted to measure to what extent the government can provide subsidies to support green technology implementation by shipping companies

4.2 (a) Environtmental analysis and compliance

An environmental benefit analysis was conducted for the MV. Labobar and MV. Gunung Dempo in order to understand to what extent the application of methanol can reduce emissions generated during the annual operation of the ships. Moreover, a compliance analysis was also performed to understand to what extent the implementation of methanol as marine fuel can satisfy future environmental regulations.

There are six emission types to be analysed, namely NOx, SOx, CO_2, CH_4, N_2O, and Particulate Matter (PM). The emission calculation is based on a one-year operation of the main engine and auxiliary engine in sailing and berthing conditions. For the case study, the basic formula to calculate the emission factor and total emission is used following the concept of IMO's 3rd GHG Study [29] and expanded as per fuel characteristics and ship operations, as follows:

$$TE = ES + EP$$

$$ES \text{ or } EP = E_{M/E} + E_{A/E}$$

$$E_{M/E} = ((\% \text{ Methanol x } EF_{Methanol}) + (\% \text{ MDO x } EF_{MDO\ M/E})) \text{ x P x t x T x LF}$$

$$E_{A/E} = EF_{MDO\ A/E} \text{ x P x t x T x n x LF}$$

$$EF_{MDO} = EF_{reference} \text{ x } SFOC_{M/E \text{ or } A/E}$$

$$EF_{Methanol} = EF_{reference} \text{ x LHV x } SFOC_{M/E}$$

Where:

TE = Total emission (tons/year)
ES = Emission during sailing (tons/year)
EP = Emission during berthing (tons/year)
EF = Emission factor (g pollutant/kWh)
P = Total operated engine power (kW)
T = Average time of sailing or berthing (hours/trip)
T = Number of trips annually
LF = Average load factor
LHV = Lower heating value (MJ/kg fuel)
SFOC = Specific fuel oil consumption (g fuel/kWh)
M/E = Main engine
A/E = Auxiliary engine

Next, one of the examples of NOx emission calculations of the main engine MV. Labobar with 80-20 fuel composition scenario, is provided:

$P_{M/E}$ = 9000 x 2 kW
$t_{sailing}$ = 198 hours/trip
$t_{berthing}$ = 0 (Main engine is off during berthing)
T = 24
LF = 0.8
LHV = 20.1 MJ/kg fuel
SFOC = 175 g fuel/kWh

$EF_{MDO} = EF_{reference} \times SFOC_{M/E}$
= 0.05684 g/gfuel x 175 gfuel/kWh
= 9.947 g/kWh

$EF_{Methanol} = EF_{reference} \times LHV \times SFOC_{M/E}$
= [0.28 g/MJ] x [20.1 MJ/1000 gfuel] x [175 gfuel/kWh]
= 0.9849 g/kWh

$E_{M/E\ sailing}$ = ((% Methanol x $EF_{Methanol}$) + (% MDO x $EF_{MDO\ M/E}$)) x P x t x T x LF
= [(80%x0.9849) + (20%x9.947) tons/kWh] x [18000 kW] x [198 h/trip] x [24 trip/year] x [0.8/1000000 g]
= 190.0486 tons/year

$TE_{NOx\ M/E}$ = 190.0486 tons/year + 0
= 190.0486 tons/year

Another emission calculation is calculated based on the above formula; both MV. Labobar and MV. Gunung Dempo are depend on various parameters that create the respective operational scenario.

Table 6 demonstrates that methanol has a clear advantage in terms of less fuel emission content compared to MDO. Even SOx, CH_4, and N_2O have zero value. Unlike the auxiliary engine of the two passenger ships, the emission factors for the main engine of MV. Labobar and MV. Gunung Dempo remains similiar since they have the same type of engine with the same SFOC, but with a different number of cylinders.

Table 6. Emission factors of MV. Labobar and MV. Gunung Dempo

Emission	Methanol (g/kWh)	MDO (g/kWh)		
		M/E	A/E Labobar	A/E Gunung Dempo
NOx	0.9849	9.947	10.7996	10.5154
SOx	0	0.462	0.5016	0.4884
CO_2	242.7075	561.05	609.14	593.11
CH_4	0	0.0105	0.0114	0.0111
N_2O	0	0.02625	0.0285	0.02775
PM	0.01512525	0.1785	0.1938	0.1887

There are four (4) significant pollutants in the internal combustion engine, particularly in a diesel engine; CO_2, NOx, SOx, and PM. The other pollutants, CH_4, and N_2O are combined with SOx and PM in Figure 6 since their value is relatively low. CO_2 and NOx are separated to create a clear picture, since their value is much higher compared to the other pollutants.

4.2 (b) N_2O, CH_4, SOx and PM analysis

According to Figure 6, N_2O and CH_4 have the lowest emissions compared to the other pollutants. The application of 80% methanol as main fuel in the main engine of MV Labobar and MV Gunung Dempo can reduce total emissions to become 59.65% compared to 100% MDO in the main engine (see Table 7). Subsequently, by increasing the composition of methanol by 5%, 4.5-4.7% of the emission reduction compared to the previous methanol percentage will be acquired. The reduction of SOx and PM is following the same trend. However, the reduction of PM was slightly lower since methanol as fuel emits some PM despite that the SOx emission factor is zero. This is because the source of PM is not only associated with the sulphur conversion during the combustion process [17].

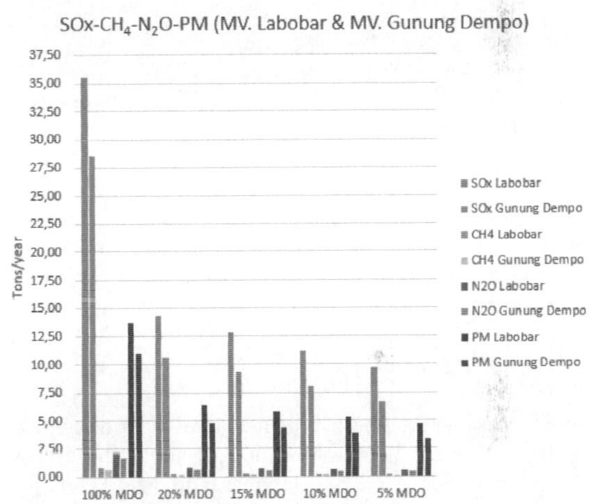

Figure 6. Total emission value of SOx-CH_4-N_2O-PM for MV. Labobar and MV. G. Dempo

Table 7. The percentage of total emission reduction of SOx-CH_4-N_2O-PM for MV. Labobar and MV. Gunung Dempo

	% reduction SOx, N_2O, CH_4		% reduction PM	
	Labobar	G.Dempo	Labobar	G.Dempo
100% MDO	-	-	-	-
M-80	59.65%	62.66%	53.62%	56.33%
M-85	64.09%	67.33%	57.69%	60.61%
M-90	68.54%	72.00%	61.76%	64.88%
M-95	72.98%	76.68%	65.83%	69.16%

Currently, Pertamina's MDO products, which are marketed in Indonesia, have complied with national and IMO regulations with a maximum sulphur content of 1.5% m/m [25]. This means that by using 100% MDO on board MV. Labobar and MV. Gunung Dempo, the ships still comply with the 3.5% m/m limit required in MARPOL Annex VI (chapter III-regulation 14).

However, according to the new regulation of SOx and PM set-up by IMO, at the beginning of January 2020 the SOx and PM limit will be 0.5% m/m. Based on Figure 7, by using 100% MDO, neither of the ships will comply with this limit. Interestingly, by using methanol as marine fuel, in all the scenarios, will help to satisfy the maximum limit of SOx and PM.

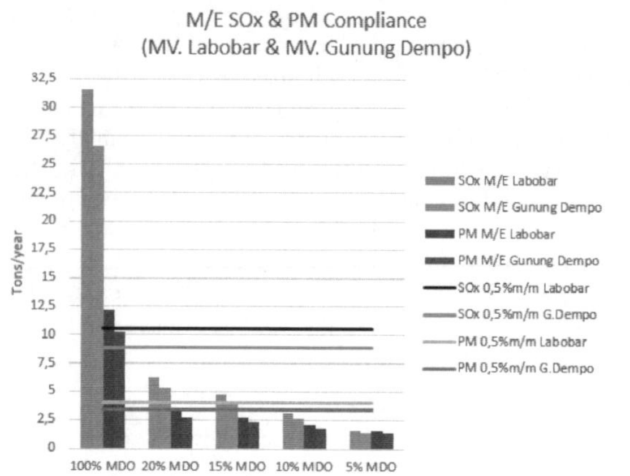

Figure 7: M/E SOx and PM value for MV. Labobar and MV. G. Dempo.

In addition to assisting in the compliance to the IMO's regulations, the application of methanol as a marine fuel can help the government commitment to protect the domestic environment from acidification, acid rain, and human health problems caused by SOx-PM pollution.

4.2 (c) NOx analysis

Table 8 presents the improvement of total NOx reduction due to methanol fuel application compared to 100% MDO application on MV. Labobar and MV. Gunung Dempo. The 50-60% reduction can be achieved just by using 80% of methanol as main fuel; the reduction will gradually increase with more methanol in the fuel composition.

Table 8. The percentage of total NOx reduction for MV. Labobar and MV. Gunung Dempo

	% reduction NOx	
	Labobar	G.Dempo
100% MDO	-	-
M-80	57.5%	60.08%
M-85	61.1%	63.84%
M-90	64.7%	67.59%
M-95	68.3%	71.35%

From Figure 9, it can be seen that both main engines of MV. Labobar and MV. Gunung Dempo fulfill the IMO NOx code Tier I that is applied for ships constructed after 1 January 2000, as they were built in 2004 and 2008. Eventhough the Tier II and III will not be imposed on these ships, an analysis for future regulation compliance can be done out of interest. For Tier II, the NOx emission value of the main engine is below the threshold for both MV. Labobar and MV. Gunung Dempo. Moreover, it will be difficult for ships of the same type and characteristic as MV. Labobar and MV. Gunung Dempo to comply with Tier III without applying measures to reduce NOx content. One way could be to implement methanol technology, considering that the use of methanol is reducing NOx emission. However, not all of the scenarios resulted in NOx emissions below the Tier III threshold (see Figure 8). The possible scenarios for the ships were M-85, M-90, and M-95.

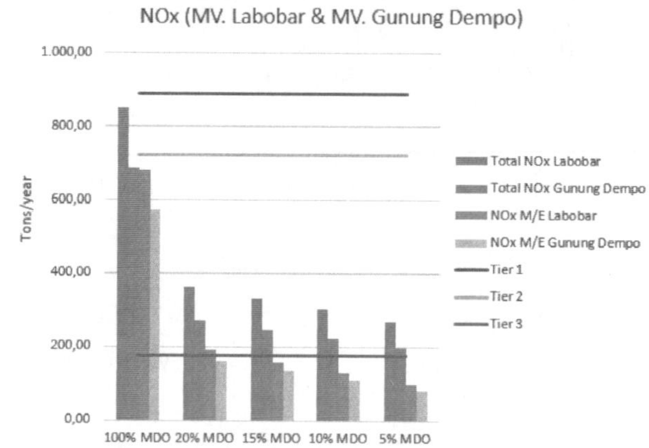

Figure 8: Emission value of NOx for MV. Labobar and MV. G.Dempo.

4.2 (d) CO_2 analysis

Total emission value of CO_2 from MV. Labobar and MV. G. Dempo can be reduced by at least 28-30%, or 12,477 tons, annually by applying M-80. The reduction will gradually increase with a higher methanol composition (see Figure 9). This reduction is possible because methanol has a lower carbon factor compared to other fuels, even LNG.

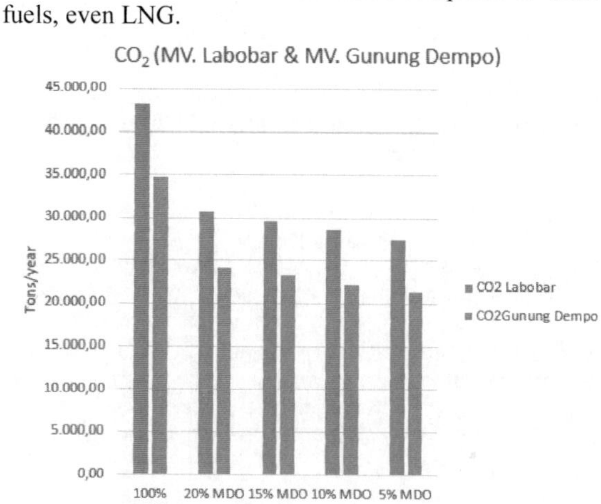

Figure 9: Total emission value of CO_2 from MV. Labobar and MV. G. Dempo.

Both of the ships are existing ships; hence, the Energy Efficiency Operational Indicator (EEOI) will be used to evaluate and quantify the energy efficiency improvement in the ship operations. Since PELNI did not have implemented yet the so-called Ship Energy Efficiency Management Plan (SEEMP), the EEOI used here is the average EEOI on an annual basis (multiplying the annual fuel consumption with the carbon factor then divided by the gross tonnage of the passenger ship and the average of the voyage annually). Table 9 shows that applying M-95 can achieve 1.54925x10-05 tonsCO_2/tonsNmiles or 36.45% of CO_2 reduction for MV. Labobar and 1.09174x10-05 tonsCO_2/tonsNmiles or 38.28% of CO_2 reduction for MV. Gunung Dempo.

Table 9. The percentage of total CO_2 reduction and EEOI for MV. Labobar and MV. Gunung Dempo.

	% reduction CO_2		EEOI (tons CO_2/tons.Nmiles)	
	Labobar	G.Dempo	Labobar	G.Dempo
100% MDO	-	-	2.43796E-05	1.769E-05
M-80	28.89%	30.33%	1.73368E-05	1.23239E-05
M-85	31.41%	32.98%	1.6722E-05	1.18551E-05
M-90	33.93%	35.63%	1.61072E-05	1.13863E-05
M-95	36.45%	38.28%	1.54925E-05	1.09174E-05

Unlike the Energy Efficiency Design Index (EEDI), EEOI is not mandatory but only recommended by the IMO. Nonetheless, the result of EEOI per individual ship will provide the overall picture to the maritime stakeholder, especially the government, on how well the CO_2 reduction effectiveness of the applied technology is performing. As shown in Figure 9 and Table 9, methanol as marine fuel is effective in reducing CO_2 and could be one of the respective government strategies to support the implementation of MARPOL annex VI chapter 4 that is already ratified by Indonesia.

4.3 MARKET-BASED INTERVENTION

Ensuring prior to establishing and stipulating a subsidy policy for methanol as a marine fuel, the government needs to have at its disposal data that demonstrates how a shipping company can improve its market share when the government interferes with subsidies. Also, the government needs to know to what extent the subsidies, in terms of quantity and condition, can interfere with the market. By an optimization approach, the government can acquire the necessary information. Moreover, the identification of which variables have the greatest influence on the policy-making on alternative fuel selection can also take place.

In this study, an optimization is conducted by using the OptQuest-Crystal Ball in the techno-economic model of MV. Labobar to achieve the above objectives. MV. Labobar model is selected as the basis of the optimization model because of the result gap between each payback time and NPV is wider than MV. Gunung Dempo. Hence, it will be easier to recognize the trend.

4.3 (a) Optimization model

There are four (4) important components that have to be identified in the optimization model: assumptions, decision variables, optimization objectives, and constraints. The explanation of these components are provided below:

a. Assumptions contain an unpredicted value or are beyond internal control [23]. In this model the assumption variable was set as follows:
   - Total Revenue is set as normal distribution with mean value according to the total revenue in 2016.
   - Engine conversion cost is set as triangular distribution with minimum cost at 250 EUR/kWh and maximum cost at 350 EUR/kWh according to the Methanol Institute report [1]. The likeliest is set up at 300 EUR/kWh based on the assumption in the techno-economy calculation.
   - The MDO price is set as normal distribution with a mean value according to the price in 2016.
   - The inflation is set as triangular distribution with minimum inflation of 4% and maximum 6% according to the regulation of the Ministry of Finance.
   - The exchange rate is set as normal distribution with a mean value at Rp. 13.319/USD.

b. Decision variables are the variables that can be controlled internally [23].
   - Percentage of methanol price compared with MDO price as the function of government subsidies. The variable is set at 43.49 as lower bounds and 73.02 as higher bounds. This value comes from the highest and lowest of the price percentage of methanol over MDO from 2004 to 2016.
   - Percentage of methanol composition is set based on the scenario in the techno-economy calculation; between 80 to 95% with interval 5%.

c. Optimization objectives are the target goal of the optimization [23]. Based on Table 4, it can clearly be observed that the boundary between recommended (light green) and not recommended (yellow) payback period is around 6-8 years. Therefore, the objective set year eight (8) as maximizing mean of payback/return rate. Maximizing the mean of payback rate will minimize the payback period.

d. Constraints are the restrictions of the decision variables [23].
   - The main engine dual fuel (methanol-MDO) will be determined as close as to when using 100% MDO. This means that the fuel cost is

close to the business-as-usual cost of the shipping company, as the minimum standard.

- The payback/ return rate after year 7 (seven) must be positive.

4.3 (b) Optimization and sensitivity analysis

From Figure 10 it is shown that the optimum decision from the government's perspective is to maintain the price of methanol to 47% or less to MDO. According to Table 2, the percentage of methanol price compared to MDO for 2016 was 59.69%. Therefore, in order to support the introduction of methanol as a green technology and a sustainable marine fuel into the market, the government needs to subsidize the methanol by 12.6%, or USD 58/tons of methanol, or Rp 610.28/liter.

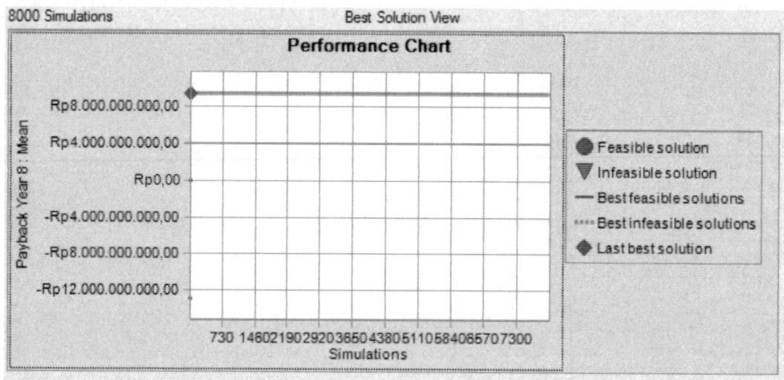

Best Solution:

Objectives	Value
Maximize the Mean of Payback Year 8	Rp9.358.244.849,12

Requirements	Value

Constraints	Left Side		Right Side
'Labobar (conversion economy)'!G7 >= 'Labobar (conversion economy)'!C7	66397860083,3...	>=	66352964545,1...
'Labobar (conversion economy)'!K28 >= 0	55960821266,523	>=	0

Decision Variables	Value
% methanol composition	0,85
% methanol price	47,00

Figure 10: OptQuest-Cristal Ball optimization result

Moreover, from this study, it can be suggested that the government should support M-85 technology in the first introduction when the market condition is as in 2016. In addition, since the methanol technology in the maritime business is relatively novel, there are opportunities to improve the technology and advancing the product. Further, with time and a massive implementation, the price will be dropped and the government subsides can be reduced.

There are also several external factors influencing a government decision to give subsidies based on the assumptions that have been made. Therefore, a sensitivity analysis was also conducted during the optimization. From figure 11, it can be concluded that the most influential external variable on government decision-making is the condition of the economic market uptake, represented in the total revenue by 52.7%. The exchange rate and the MDO price had almost the same influence, 25.2%, and 21.9% respectively. However, the engine conversion cost and the inflation had a smaller effect.

In general, the result of the sensitivity analysis is logically acceptable since the government does not want to impose any new technology that can disrupt the maritime industry, for instance an exorbitant price of technology or a sluggish maritime business market. The MDO price is also considered as an external factor that can change the government decision. For instance, when the price of MDO becomes higher, the government will try to find a solution to maintain its maritime business such as introducing alternative fuels or even subsidizing the use of certain fuels.

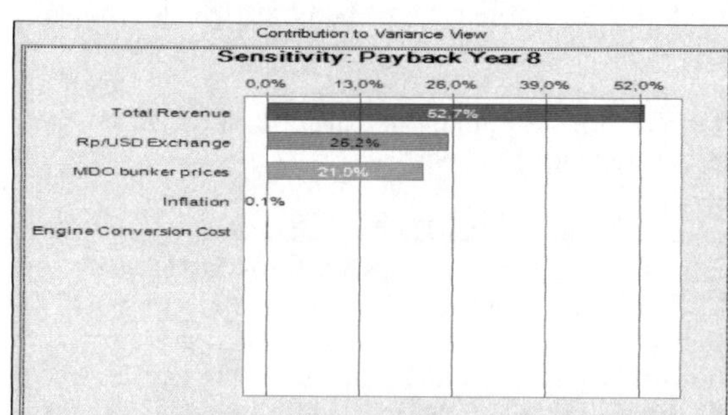

Figure 11: Sensitivity analysis of external factor for payback year-8

4.3 (c) Verification

Moreover, verification of the optimization result is important. A simple verification was made by manually calculating the fuel cost of the main engine of MV. Labobar for each methanol composition scenario. Figure 12 demonstrates that the position of the 47% line was the same as the 100% MDO line; this means that the fuel cost of the methanol as a main fuel with various compositions will be close to the fuel cost of a business-as-usual condition of the shipping company. From the government perspective, 47% was the optimum price of methanol where the optimum subsidies can maintain the market and keep the shipping company making profits as usual.

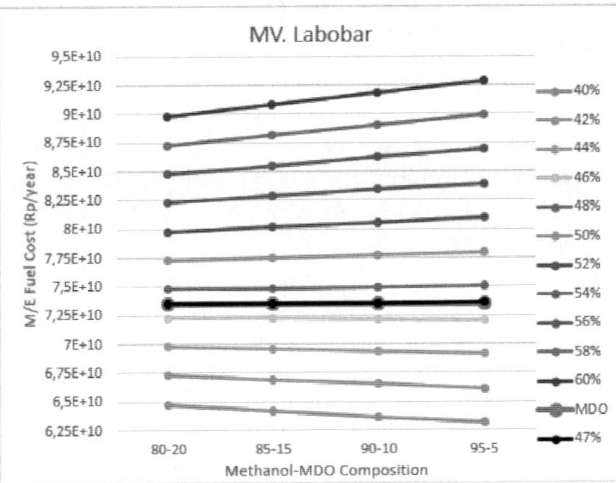

Figure 12: Verification of the optimization result based on fuel cost

From the ship owner's perspective, the percentage was the minimum price of methanol to decide on investment of methanol technology on the fleet, with a favourable payback period. By inputting the scenario in methanol price to 47% with a composition of 85-15 in the techno-economic calculation, some economic criteria could be then defined. The payback period can be achieved in 7 years and 3 months, with the positive NPV Rp86,057,237,977.600. When the criteria are plotted to Table 4, then the position will be between the light green and yellow area; this eans that these criteria are the minimum for the shipping companies to maintain their profit in the business-as-usual scenario and in the case the application of methanol is introduced.

4.4 DISCUSSION

"Competitiveness" of methanol as a marine fuel generally depends on ship productivity and the price differences between methanol and MDO. Considering the different results of techno-economic calculations in the feasibility investment of MV. Labobar and MV. Gunung Dempo, this means that the model should be applied separately to each individual ship. The result can not be generalized as the reference for other passenger ships, since it may vary depending on the revenue gained (profit), engine size, maintenance cost, cargo capacity, and remaining economic life of each passenger ship.

However, the trend of the combinatorial scenario analysis can be considered for both ships as the reference for other passenger ships, since there are similar trends and an interesting relationship between NPV, payback period, percentage of price, and percentage of methanol price to MDO. The payback period and NPV for each percentage of methanol composition are sensitive to the percentage of the methanol price. Up to 46% of the methanol price to MDO, the improvement of payback period and NPV are directly proportional to the increment of methanol composition as main fuel. At the same time, around 48% above the trend, the situation will be the opposite. These results can be considered as the indicative strategy for ship-owners to select the "right" operational option when dealing with the current market situation. When the price of methanol is close to or above 48% of the MDO price, then the lowest set-up methanol composition (80% methanol-20% MDO) can be operated to maintain the profit and payback time.

In terms of regulation compliance, running with dual-fuel methanol propulsion significantly reduces related emissions. Generally, as seen from Tables 6-9, the higher methanol composition used as a main fuel, the higher emission reduction is achieved. Further, most of the scenario can comply with the recent and upcoming regulations, particularly MARPOL Annex VI. Therefore, from the policy compliance point of view, the application of methanol as a marine fuel is feasible to get government support since it will help the government commitment to protect the domestic environment from negative impacts to the environment and human health caused by pollution from ships.

Decision-making and policy analysis using optimization can be performed as one of the government approaches in determining the optimum point and condition to introduce and establish methanol as a marine fuel. The optimum point that the government should maintain is the methanol price at 47%. The MDO price has a similar trend in the combinatorial scenario analysis from the ship-owner's perspective. Moreover, from the sensitivity analysis result, it is evident that here are three main external variables that have to be taken into account in the policy-making: market situation, methanol price and exchange rate. The complete work of this research can be found in the first author's MSc dissertation conducted at the World Maritime University [36].

5. CONCLUSIONS

Domestic ships are one of the most important means of transport to connect the numerous islands in an archipelagic country, like Indonesia. A certain number of the domestic ships, particularly passenger ships are assigned to deliver services in uncompetitive commercial areas. Unfortunately, the vast majority of the current passenger ships are dependent on fossil fuel, and therefore vulnerable to fluctuations of fuel oil prices. In order to promote a sustainable and more "environmental friendly" shipping market, methanol fuel can be a very suitable for Indonesia. Abundant potential feedstock with availability of methanol producers and existence of relevant infrastructure in the country under discussion have all been identified within the context of the analysis in hand. As already highlighted, running with dual-fuel methanol propulsion significantly reduces ship's emissions to air. In the emission calculations previously performed, most of the scenarios can comply both with the recent and upcoming regulations, particularly with those of MARPOL's Annex VI.

Subsequently, an economic analysis was performed by using a techno-economic model based on case studies of two passenger ships owned by PELNI: MV. Labobar and MV. Gunung Dempo. The combinatorial scenario approach, which is the combination of economic measures of merit (NPV and payback period) with the technical solution scenario (main-pilot fuel set up), effectively provides a broader overview for ship-owners not only to determine the feasibility of the investment in methanol technologies, but also to determine which ships are "eligible" for retrofitting and what scenarios of engine set-up should be operated on-board the ship based on ship age, ship productivity, as well as current and long-term market conditions.

It was found that the competitiveness of methanol application is mainly dependent on ship productivity and the price differences between methanol and MDO. Productivity of passenger ships was represented with revenue; it can be improved by modifying and improving container cargo capacity (MV, Labobar is a 2-in-1 ship, passenger and container cargo), or by introduction of "green technology" subsidies as a market-based intervention by the government.

However, there is a trade-off situation in the market-based intervention. Ship-owners tend to get high income by collecting as many incentives as possible, while the government needs to provide subsidies that are based on a limited state budget. Therefore, an optimization approach was developed and performed by utilizing the combinatorial scenario model; hence, the optimum methanol price was evaluated. The optimization result revealed that the optimum price of methanol was when the percentage of methanol price compared to MDO was 47%. That is the optimum percentage where both fuels costs are at the same value. Moreover, it could be the reference for the government to target a percentage by providing subsidies for the market to be maintained and to help the shipping company maintain profits as if they had been operating their vessel with MDO.

Providing support for methanol as a marine fuel could increase domestic methanol production and also encourage its use in other industrial sectors. Methanol hopefully can fulfil the energy transition needed since the oil reserves in Indonesia are for the time being steadily decreasing. However, there are several issues that must be addressed and considered:

- In the short term, an initiative should come first from the government with a national policy including financial support, such as subsidies to allow the stakeholders to develop sustainable energy strategies ranging from model to full-scale experiment. This is important for gaining the trust of those shipping companies that do not want to take the "unnecessary" risk. The government also needs to stimulate further research on this issue by academic institutions, engine manufacturers, methanol producers, and various other parties involved into developing the market for methanol as a marine fuel. Moreover, in order to bring clarity regarding the legal basis, the government should work together with classification societies to develop safety regulations for domestic passenger ships running on methanol.
- In the medium term, the government should develop a rigid energy policy, as well as a national strategic roadmap that includes methanol as one of the alternative fuels for transportation, particularly marine transport needs. The policy and strategic roadmap need to consider an incentives scheme, allocation of methanol fuel supplies, an inter-ministerial coordination framework, and explicit responsibilities for each party involved. In addition, Indonesia still has abundant resources of coal and natural gas that have not yet been absorbed by the domestic and international market. Therefore, it would be favourable to increase methanol production using coal and natural gas in the medium term of the energy transition.
- In the long term, considering Indonesia has ample waste as renewable feedstock resources, such as plantation waste and municipal waste, it would be favourable to shift the methanol feedstock from natural gas and coal into the more sustainable feedstock. Further, utilization of these resources could help the environment by reducing emissions and even by creating a circular economy.

8. REFERENCES

1. Andersson, K and Salazar, C.M. *Methanol as a Marine fuel report*. London, UK: FCBI energy. (2015).

2. Aronietis, R., Sys, C., Vanelslander, T. *Ship retrofit solutions: economic, energy and environmental impacts, in: Ehlers, S. et al.* Maritime-port technology and development. pp. 57-66. (2014).

3. Bank of Indonesia Database. Retrieved July 15, 2017, from http://www.bi.go.id/id/moneter/informasi-kurs/transaksi-bi/Default.aspx. (2017).

4. Batidzirai, B., Smeets, E. M. W., & Faaij, A. P. C. *Harmonising bioenergy resource potentials—Methodological lessons from review of state of the art bioenergy potential assessments*. Renewable and Sustainable Energy Reviews, 16(9), 6598-6630. (2012).

5. BKI database. Retrieved July 15, 2017, from http://ogs.bki.co.id/. (2017).

6. Bozzano, G., & Manenti, F. *Efficient methanol synthesis: perspectives, technologies and optimization strategies*. Progress in Energy and Combustion Science, 56, 71-105. (2016).

7. Brigham, E. F., & Ehrhardt, M. C. *Financial management: Theory & practice (thirteen edition)*. Mason, USA: Cengage Learning. (2011).

8. Bromberg, L., & Cheng, W. K. *Methanol as an alternative transportation fuel in the US: Options for sustainable and/or energy-secure transportation*. Cambridge, MA: Sloan Automotive Laboratory, Massachusetts Institute of Technology. (2010).

9. Buhaug, Ø., Corbett, J. J., Endresen, Ø., Eyring, V., Faber, J., Hanayama, S., ... Yoshida, K. *Second IMO GHG study 2009*. London, UK: IMO. (2009).

10. Bunkerindex database. Retrieved July 15, 2017, from http://www.bunkerindex.com/ prices/. (2017).

11. Clarkson Intelligence network. Retrieved July 15, 2017, from http://sin.clarksons.net/. (2017).

12. Coordinating Ministry of Maritime Affairs (Menkomar). *Tugas dan Fungsi (Duties and functions)*. Retrieved from https://maritim.go.id/. (2015).

13. Directorate of Upstream Chemical Industry (DIKH). *Hilirisasi pembangunan industri berbasis migas dan batubara (Downstreamization industrial development based on oil and Gas and coal)*. Presentation material from work program of Ministry of Industry Republic of Indonesia. Jakarta, Indonesia: Author. (2016).

14. ESDM. *Handbook of Energy & Economic Statistics of Indonesia 2016*. Jakarta, Indonesia: Author. (2016).

15. Geological Agency of Indonesia (BGI). *Executive Summary Pemutakhiran Data dan Neraca Sumber Daya Energi 2015 (Executive Summary Data Updates and Energy Balance Sheets 2015)*. Jakarta, Indonesia: ESDM. (2015).

16. Hasan, M. H., Mahlia, T. M. I., & Nur, H. *A review on energy scenario and sustainable energy in Indonesia*. Renewable and Sustainable Energy Reviews, 16(4), 2316-2328. (2012).

17. IMO. *Methanol as marine fuel: Environmental benefits, technology readiness, and economic feasibility*. London, UK: Author. (2016).

18. Kaltim Methanol Industry (KMI). *History-manufacturing*. Retrieved from http://kaltimmethanol.com. (2015).

19. Laakso, T. Email correspondence (Wartsila). (2017).

20. Methanol Institute. *Methanol safe handling manual 4th edition*. California, USA: Author. (2017).

21. Nizami, A. S., Rehan, M., Waqas, M., Naqvi, M., Ouda, O. K. M., Shahzad, K., ... & Pant, D. *Waste Biorefineries: Enabling Circular Economies in Developing Countries*. Bioresource Technology. (2017).

22. Olah, G. A., Goeppert, A., & Prakash, G. S. *Beyond oil and gas: the methanol economy*. Weinheim, German: WILEY-VCH Verlag GmbH & Co. (2006).

23. Oracle. *Crystall ball user's guide release 11.1.2.3*. (2013).

24. PELNI. *Anual Report 2016*. Retrieved from https://www.pelni.co.id. (2016).

25. Pertamina. *Spesifikasi bahan bakar minyak diesel (Specification of diesel oil fuel)*. Retrieved from http://www.pertamina.com/industrialfuel/media/20709/ minyak-diesel.pdf. (2009).

26. Prasodjo, E., Nurzaman, H., Walujanto., Rosdiana., Ismutadi, P., Malik, C., ... Sauqi, A. *Outlook energi indonesia 2016 (Indonesia energy outlook 2016)*. Jakarta, Indonesia: National Energy Council (DEN). (2016).

27. Santoso, T. Email correspondence (Pelni). (2017).

28. Shen, W., Han, W., Chock, D., Chai, Q., & Zhang, A. *Well-to-wheels life-cycle analysis of alternative fuels and vehicle technologies in*

China. Energy Policy, 49, 296-307. DOI: 10.1016/j.enpol.2012.06.038. (2012).

29. Smith, T. W. P., Jalkanen, J. P., Anderson, B. A., Corbett, J. J., Faber, J., Hanayama, S., ... & Raucci, C. *Third imo ghg study 2014*. London, UK: IMO. (2014).

30. Srivastava, A. *Compliant strategy for shipowners towards sustainable maritime transport: A decision framework for air emission reduction measures (Master Dissertation)*. Malmo, Sweden: World Maritime University. (2016).

31. Stojcevski, T. *Methanol as engine fuel, status Stena Germanica and market overview*. Qatar: Middle East Methanol Forum. (2014).

32. Sugiyono, A., Boedoyo, M.S., Fitriana, I., Niode, N., Siregar, E., Paminto, A.K., ... Suryani. *Outlook energi indonesia 2016: pengembangan energi untuk mendukung industri hijau (Indonesia energy outlook 2016: energy development in supporting green industry)*. Jakarta, Indonesia: BPPT. (2016).

33. UNEP. *Reforming energy subsidies: opportunities to contribute to the climate change agenda*. United Nations Environment Programme Division of Technology, Industry and Economics. (2008).

34. Winrock. *Buku panduan konversi POME menjadi biogas: pengembangan proyek di Indonesia (Guideline for conversion project of POME to biogas: project development in Indonesia)*. Jakarta, Indonesia: United State Agency for International Development (USAID). (2015).

35. Zhen, X., & Wang, Y. An overview of methanol as an internal combustion engine fuel. *Renewable and Sustainable Energy Reviews*, *52*, 477-493. (2015).

36. Priyanto, E.M. *A feasibility study of methanol fuelled domestic passenger ships in Indonesia (Master Dissertation)*. Malmo, Sweden: World Maritime University. (2017).

9. AUTHORS BIOGRAPHY

Eko Maja Priyanto holds the position of Senior Inspectore at SBU Marine and Offshore - PT. Biro Klasifikasi Indonesia (Persero). He is responsible for supervising several projects in marine and offshore unit. His previous experience includes mechanical engineering of dual fuel engine, technical advisor of the implementation of fuel monitoring system, developing classification society regulation related to dual fuel engines.

Dr. Aykut I Ölçer is a naval architect and marine engineer holding the position of Nippon Foundation Professorial Chair in Marine Technology and Innovation at the World Maritime University (WMU), Malmö, Sweden. He currently leads the WMU Maritime Energy Research Group (http://wmu.se/research) as well as serves as the Head of Maritime Energy Management Specialisation at WMU. At the same time, he is the Editor in Chief of WMU Journal of Maritime Affairs (JOMA) and Book Series.

Dr. Dimitrios Dalaklis holds the position of Associate Professor (Safety and Security) at the World Maritime University (WMU), in Malmo-Sweden. His expertise revolves around the extended Maritime Education and Training (MET) domain, as well as maritime safety and security issues, with emphasis on LNG related topics.

Dr. Fabio Ballini holds the current position of Assistant Professor (Maritime Energy Management) at the World Maritime University (WMU), in Malmo-Sweden. He is dealing with research topics related to Port Energy Management Planning, Gaseous Emissions from Ships in Harbors, Externality Cost in Transportation, Externality Modeling, Clean-Tech Solutions and Energy Audit Systems in Port.

TOWARDS ELECTRIFICATION OF RO-RO PASSENGER FLEET IN THE ADRIATIC SEA

M Perčić, I Ančić and **N Vladimir**, University of Zagreb, Faculty of Mechanical Engineering and Naval Architecture, Croatia

SUMMARY

With the recent introduction of new requirements for ships energy efficiency, as well as continuous rise of fuel price, there is a growing interest by many shipyards and shipowners to reduce fuel oil consumption. This is particularly important for ro-ro passenger ships in the Adriatic Sea. Various innovative energy efficient technologies could be implemented for these ships in order to reduce their environmental footprint. In this paper the viability to electrify these ships is analysed. This was done by analysing the energy efficiency of the current fleet, including the estimation of the fuel oil consumption and the analysis of ferry routes in the Adriatic Sea, as well as the number of passengers and vehicles transported. Based on this analysis, it is concluded that the electrification of the ro-ro passenger fleet in the Adriatic Sea could be considered as a viable option in the future.

NOMENCLATURE

EEDI	Energy Efficiency Design Index ($g\ CO_2\ t^{-1}\ nm^{-1}$)
FC_{annual}	Total annual fuel oil consumption on every route (t)
FC_{route}	Average annual fuel oil consumption on a route (t)
FC_{trip}	Fuel oil consumption for one trip on a route (t)
GT	Gross tonnage
IMO	International Maritime Organization
MARPOL	The International Convention for the Prevention of Pollution from Ships
MEPC	Marine Environment Protection Committee
N_{route}	Number of trips on a route per year
P_P	Average propulsion power (kW)
P_T	Average total power demand (kW)
R^2	Coefficient of correlation
SFOC	Specific fuel oil consumption ($g\ kW^{-1}\ h^{-1}$)
WROS	World Register of Ships
a	Parameter in the regression curve (-)
c	Parameter in the regression curve (-)
t_{trip}	Trip duration (h)
v	Ship speed (kn)

1. INTRODUCTION

In the recent years there is a growing interest by many shipyards and shipowners to introduce energy efficient technologies on-board new and existing ships. The primary aim of these technologies is to save fuel oil. There are two main reasons for such actions: introduction of new chapter 4 within MARPOL Annex VI [1], including new requirements for ships energy efficiency, as well as continuous rise of fuel oil price. Even though the primary targets of the new regulation introduced by MEPC are large oceangoing vessels, the incentive to increase energy efficiency is also strong for ships engaged in short-sea shipping. This particularly refers to ro-ro passenger fleet. Ro-ro passenger ships operate in shorter routes, spend much more time in ports manoeuvring or at berth, and hence have a higher negative impact on the port environment [2]. Also, these ships transport passengers and cars, so there is a significant public pressure through various public campaigns to make these ships "greener", i.e. more environmentally friendly.

Regulations within MAROPL now require for every ship of GT 400 and above to have an International Energy Efficiency (IEE) Certificate on-board. In order to obtain it, a new ship has to comply with the Energy Efficiency Design Index (EEDI) requirement, i.e. for every new ship the Attained EEDI has to be calculated (according to the prescribed methodology) and should not be higher the Required EEDI. The Required EEDI is defined based on the ship type, size and the date of building, meaning ships in the future will have to be more energy efficient compared to today's ships [3].

The aim of this research is to explore options for electrifying ro-ro passenger fleet in the Adriatic Sea. As a basis for that kind of activity it is necessary first to analyse operational profile of target vessels. Similar study was performed for ships operating in Norwegian waters by Jafarzadeh and Schjølberg [4], where they confirmed relatively similar operating profiles of ships engaged in short-sea shipping, but very distinct of those operating far from shore. Moreover, among the considered ship types, they report that reefers spend more operational time close to the diesel engine design condition, while, offshore and passenger ships show the most dynamic operational profiles and spend a large percentage of their operational time at a partial load, away from diesel engine design conditions. Jafarzadeh and Schjølberg [4] claim that such ships can benefit from hybridisation, diesel-electric propulsion, and other electric concepts, such as batteries and fuel cells. Irrespective on the electricity origin on-board, electric propulsion nowadays represents very important and attractive research topic due to both environmental issues and possible money savings. Namely, it is known that

hybrid technology in ships can significantly reduce fuel oil consumption and local CO_2 emissions. Kalikatzarakis et al. [5] analysed a tugboat powered by a hybrid propulsion plant with power supply that can be recharged with renewable shore power. That hybrid configuration has the additional challenge to determine the optimal power-split between three or more different power sources, in real-time, and to optimally deplete the battery packs over the mission profile. Jeong et al. [6] introduced new decision-making process which is used to compare the performance of a ship with either diesel electric hybrid propulsion or conventional propulsion systems. A case study was carried out to compare the performance of both propulsions from cost, environmental and risk perspectives. The key process in their research was to convert all incomparable values into monetary values, thereby enabling the impacts of each criterion to be compared and integrated in a straightforward manner. Their results indicate that the use of hybrid propulsion has advantages over both diesel electric and diesel mechanical propulsion systems for a set of operating scenarios. Motivated by the extensive exploitation of electric power in ships, Kanellos et al. [7] proposed an optimal power management method for ship electric power systems comprising integrated full electric propulsion, energy storage and shore power supply facility. Beside general economic aspects inherent to all-electric ships, special problem represent load fluctuations [8], and there are number of recent works dealing with this issue. Rodrigues et al. [9] investigated impact of electric propulsion on the electric power quality of vessels, particularly addressing a significant increase of non-linear loads in the system, due to variable frequency drives feeding the motor. This important issue is also studied by Hou et al. [10] who explored a new solution, namely a combined battery and flywheel (B/FW) hybrid energy storage system (HESS) as a buffer to isolate load fluctuations from the shipboard network.
As indicated above, this paper represents basic research towards electrification of ro-ro passenger vessels operating in the Adriatic Sea and has the aim to determine the current status of ro-ro passenger fleet in the Adriatic Sea. Moreover, it is limited to fleet operating on the Croatian coast of the Adriatic Sea. There are several reasons for that. The Croatian side of the Adriatic Sea has numerous islands and indented coastline, so there are many more ferry lines connecting land and islands, as well as islands with each other, compared to other countries in the Adriatic Sea. Also, the importance of these lines in Croatia is very high, not only for local population, but as well as for tourists visiting Croatia in large numbers, especially during the summer period. Due to tourism, the sensitivity of the marine environment in Croatia is also increased, since any pollution can have devastating effect not only on the environment, but also on the entire economy.

In the next section, the methods used to determine current status of ro-ro passenger fleet in the Adriatic Sea are presented. Further, the results obtained using these methods are presented and discussed, and finally, the conclusions are drawn and guidelines for further research are outlined.

2. METHODS

In order to evaluate the viability of electrifying ro-ro passenger fleet in the Adriatic Sea, first the status of a current fleet should be reviewed. As indicated in the introduction section, two main incentives to improve energy efficiency of these ships are regulations and fuel oil price. So it is necessary first to determine the energy efficiency performance of the current fleet, and second to determine their total fuel oil consumption.
The energy efficiency analysis is based on data available from IHS Fairplay's "World Register of Ships (WROS)" database [11] using the approach adopted for the EEDI calculation, as well as studies available in the literature [12, 13].
The total benefit for the society, i.e. the economic output for ro-ro passenger ships is expressed in number of passengers and vehicles transported. These data are obtained from overviews annually presented by the Croatian Agency for short-sea liner shipping [14].
There are two methods to determine total annual fuel oil consumption for ro-ro passenger ships in the Adriatic Sea. First, a very direct method is to gather data on how much fuel oil did the ship operators spend in a year. But, precisely this data is not publicly available.
The other method is to sum up the annual fuel oil consumed on every route. To calculate the average fuel oil consumption on a route FC_{route}, fuel oil consumption for one trip on every route FC_{trip} should be multiplied by the number of trips on that route N_{trips} in a year:

$$FC_{annual} = \sum_{i=1}^{n} FC_{route,i} \quad (1)$$

$$FC_{route} = FC_{trip} \cdot N_{trips} \quad (2)$$

FC_{trip} can be calculated based on the average total power demand P_T, specific fuel oil consumption ($SFOC$) and route duration t_{trip}:

$$FC_{trip} = P_T \cdot SFOC \cdot t_{trip} \quad (3)$$

Values of P_T and $SFOC$ depend on many parameters, such as ship, route, weather, loading conditions etc. Since the ship operators alter different ships on different routes at a random level it is very hard to determine precisely the fuel oil consumption on a specific route. But it is relatively easy to determine the average speed on a route since the time schedules do not change with a ship. If a faster ship (with higher installed power) is used for the same route as the slower one, it will simply reduce its engine power and sail at the reduced speed. It is assumed that while operating at this reduced speed, the total power it uses will be similar as for the slower ship, i.e. it is assumed the propulsion power used for a route is dependent on ship speed, regardless of the ship used on that route. A regression curve linking propulsion power P_P and ship speed v can be determined as a power function, with a and c as parameters:

$$P_P = a \cdot v^c \qquad (4)$$

where parameter c should equal around 3. These parameters are determined based on the data from ro-ro passenger ships in the WROS database [11]. To take into account the auxiliary power demand, the propulsion power P_P is increased by 10% to obtain total power P_T. This assumption will definitely cause some inaccuracy, but the only alternative is to monitor every ships power and speed for a year on every route, and such data are not currently available.

SFOC is determined depending on the engine speed, as proposed by [13], i.e. it is assumed for medium speed engines the *SFOC* is 180 g/kWh, while for high speed engines the *SFOC* yields 215 g/kWh.

3. RESULTS

3.1 ENERGY EFFICIENCY ANALYSIS

The analysis revealed that the majority of 44 ro-ro passenger ships operating on the Croatian side of the Adriatic Sea are outdated. The average ship age is 26 years, with only 6 ships under 10 years of age. The energy efficiency performance of a typical ro-ro passenger ship for the Adriatic Sea was already performed in [13]. It revealed that this ship does not comply with the EEDI requirements, i.e. its Attained EEDI is 83.41 g CO_2/t·nm which is notably higher than the EEDI reference line value for this ship type and size of 71.57 g CO_2/t·nm.

In total there are 27 ferry lines operating in the Adriatic Sea analysed in this paper. 21 lines connect the Croatia mainland and islands, 1 line connects two parts of Croatia mainland (Ploče - Trpanj), and there are 2 lines connecting islands with each other (Merag - Valbiska and Lopar - Valbiska), as well as 3 international lines connecting Croatia and Italy, Figure 1.

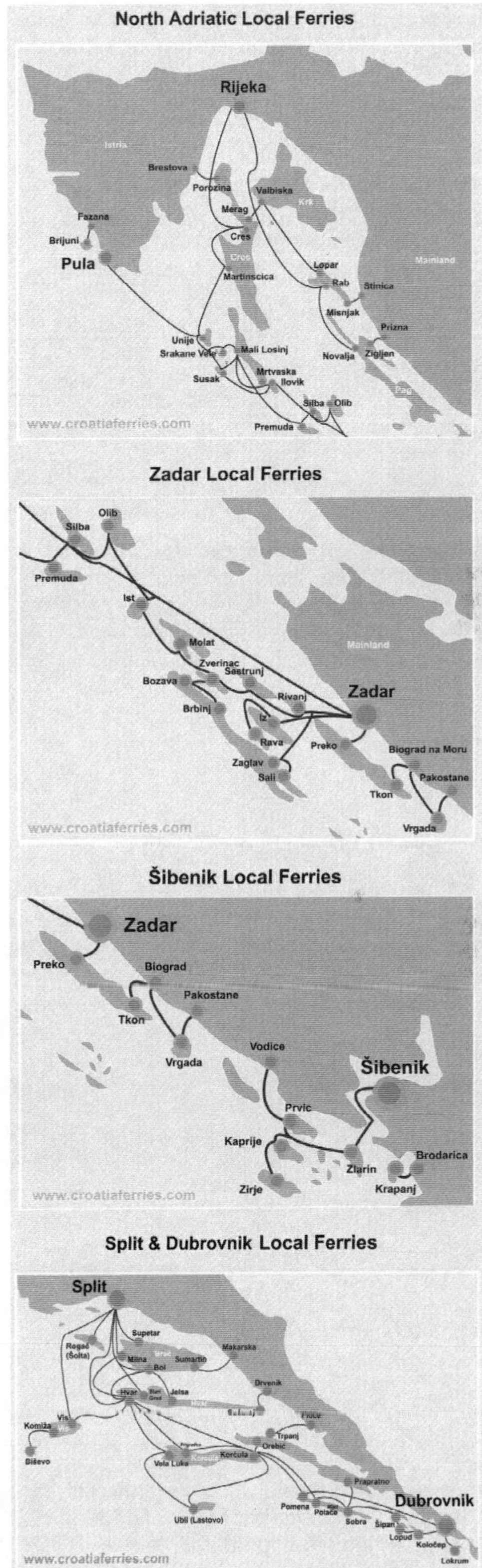

Figure 1. Croatia ferry map [15]

In total there are around 50,000 trips taken every year by ferries during which they sail in total around 400,000 nautical miles. The data on the number of passengers and vehicles transported on the selected routes is presented in Table 1, according to [14].

Table 1. Number of passengers and vehicles transported by ferries in Croatia [14]

Year	Total number of passengers	Total number of vehicles
2017	10,899,630	3,294,172
2016	10,236,946	3,102,264
2015	9,851,454	2,993,793
2014	9,350,276	2,825,886
2013	9,271,372	2,771,168
2012	9,149,478	2,764,073

3.2 TOTAL FUEL OIL CONSUMPTION

Based on the data in the WROS database a regression curve between the ship speed and engine power is calculated. In total, there are 44 ro-ro passenger ships operating in the Adriatic Sea on the selected routes. For 10 of these ships data on the ship speed is missing, so the regression analysis is performed for 34 ships, Figure 2.

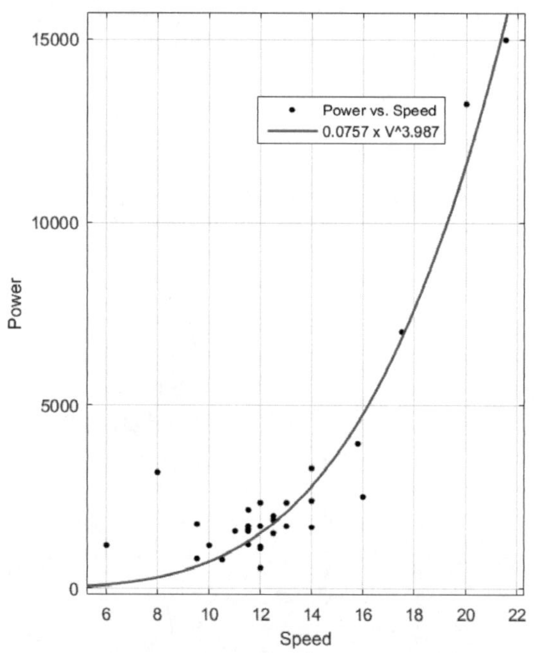

Figure 2. Regression curve $P_P = f(v)$

The analysis showed that the main engine power is dependent on ship speed roughly on the 4th potential, i.e. parameter a equals 0.0757, and parameter c equals 3.987, with relatively high correlation factor which equals $R^2 = 0.91$.

Using methodology described in the previous section the total annual fuel oil consumption for ro-ro passenger ship fleet on the selected routes in the Adriatic Sea is around 9,100 tonnes.

4. DISCUSSION

As stated previously, there are two main incentives to electrify ro-ro passenger fleet in the Adriatic Sea. The first is the regulations recently adopted by the IMO. The analysis in this paper revealed that the majority of fleet is outdated. Furthermore, the analysis performed by Ancic et al. [13] revealed that a typical ro-ro passenger ship used in the Adriatic Sea has poor energy efficiency performance since its Attained EEDI is higher than the EEDI reference line value. In the Phase 3 of the EEDI application the Required EEDI will have to be 30% lower than that reference line value. This means that the Attained EEDI for new ships have to be at least 40% lower compared to an existing ship, i.e. new ships will have to reduce their fuel oil consumption by roughly 40% (assuming their operational characteristics remain the same). Annually this reduction in the fuel oil consumption would be around 3,600 tonnes, i.e. it would lead to the reduction of operational costs by around 3.4 million Euros. This opens possibilities to implement innovative energy efficient technologies. So it is reasonable to say that the electrification of new ro-ro passenger ships could be a viable option in order to satisfy the future regulations.

The second incentive is the fuel oil price. As shown in the previous paragraph, a 40% fuel oil consumption reduction is a requirement for new ships from 2025. The total annual fuel oil consumption for ro-ro passenger ships in the Adriatic Sea on the selected routes is around 9,100 tonnes. This value is an estimate based on the average main and auxiliary engine powers, SFOC, route duration, number of trips per route per year and total number of routes. Since ships are altered between different routes at a random interval, it is not possible to determine exactly the fuel oil consumption for a specific route in advance. Hence, the engine power is determined based on an average speed on a route as a power curve with parameters a and c. In this relation the parameter c equals 3.987, which is higher than expected. It has to be noted that the engine power is dependent on various parameters, such as ship speed, ship particulars, hull form, propulsion system, weather conditions etc. In the ro-ro passenger fleet observed, the bigger ships have somewhat higher design speed. Since the ship size is not taken into account, this led to the increase of the parameter c. But a relatively high correlation factor confirms that this regression curve approximates this relation relatively well.

The fleet analysis revealed that 9 ro-ro passenger ships have medium speed engines, while the others have high speed engines. Since these ships with medium speed engines are commonly used for longer international routes, it is assumed the specific fuel oil consumption for these routes is 180 g/kWh, while for other routes the SFOC is assumed to be 215 g/kWh.

When observing the ro-ro passenger fleet economic output, i.e. the number of passengers and vehicles transported, a steady increase can be noticed. In a five year interval, from 2012 to 2017, the increase in the number of passengers, as well as in the number of vehicles is around 20%. In part, this increase was compensated with the introduction of additional lines on some routes (which led to the increase of fuel oil consumption and hence the environmental impact). But also, many lines now transport more passengers per route (many ships no longer operate half empty). This has negligible impact on the fuel oil consumption, but high on the ship income. So this potential increase in income also opens possibilities to implement new innovative energy efficient technologies, where electrification could again be considered as a viable option.

5. CONCLUSION

In this paper two main incentives to improve energy efficiency of a ro-ro passenger fleet in the Adriatic Sea are investigated. The first incentive is regulatory, coming from IMO as a requirement to improve energy efficiency of new ships. Since the energy efficiency performance of the current ships is rather poor, new ships will have to be significantly better, i.e. new ships from 2025 will have to reduce their fuel oil consumption by around 40% compared to the current fleet.

The second incentive is economical, i.e. the ship owners want to reduce their expenses. Considering the increase in the number of passengers and vehicles transported, as well as the need to increase the energy efficiency, it can be concluded that the electrification of ro-ro passenger ship fleet in the Adriatic Sea could definitely be a viable option in the future. In order to confirm this, a techno-economic analysis should be performed for each route in order to determine the optimal configuration of the ship power system.

6. ACKNOWLEDGEMENTS

This research was supported by the Croatian Science Foundation under the project Green Modular Passenger Vessel for Mediterranean (GRiMM), (Project No. UIP-2017-05-1253). Miss Maja Perčić, Ph.D. student is supported through the "Young researchers' career development project – training of doctoral students" of the Croatian Science Foundation, funded by the European Union from the European Social Fund.

7. REFERENCES

1 MEPC, 'Resolution 203(62)', 2011.

2 RUNKO LUTTENBERGER, L., ANČIĆ, I., ŠESTAN, A., 'The viability of short-sea shipping in Croatia', *Brodogradnja, 64, 472-481*, 2013.

3 MEPC, 'Resolution 245(66)', 2014.

4 JAFARZADEH, S., SCHJØLBERG, I., 'Operational profiles of ships in Norwegian waters: An activity-based approach to assess the benefits of hybrid and electric propulsion', *Transportation Research Part D: Transport and Environment, 65, 500-523,* 2018.

5 KALIKATZARAKIS, M., GEERTSMA, R.D., BOONEN, E.J., VISSER, K., NEGENBORN, R.R., 'Ship energy management for hybrid propulsion and power supply with shore charging', *Control Engineering Practice, 76, 133-154,* 2018.

6 JEONG, B., OGUZ, E., WANG, H., ZHOU, P., 'Multi-criteria decision-making for marine propulsion: Hybrid, diesel electric and diesel mechanical systems from cost-environment-risk perspectives', *Applied Energy, 230, 1065-1081,* 2018.

7 KANELLOS, F., ANVARI-MOGHADDAM, GUERRERO, J.M., 'A cost-effective and emission-aware power management systems for ships with integrated full electric propulsion', *Electric Power Systems Research, 150, 63-75,* 2017.

8 HOU, J., SUN, J., HOFMANN, H., 'Adaptive model predictive control with propulsion load estimation and prediction for all-electric ship energy management', *Energy, 150, 877-889,* 2018.

9 RODRIGUES, T.A., NEVES, G.S., GOUVEIA, L.C.S., ABI-RAMIA JR., M.A., FORTES, M.Z., GOMES JR., S., 'Impact of electric propulsion on the electric power quality of vessels', *Electric Power Systems Research, 155, 350-362,* 2018.

10 HOU, J., SUN, J., HOFMANN, H., 'Control development and performance evaluation for battery/flywheel hybrid energy storage solutions to mitigate load fluctuations in all-electric ship propulsion systems', *Applied Energy, 212, 919-930,* 2018.

11 IHS FAIRPLAY, 'World Register of Ships Manual', 2013.

12 ANČIĆ, I., ŠESTAN, A., VLADIMIR, N., KLISARIĆ, V., 'Influence of New Power Sources on the Attained EEDI', *Influence of EEDI on Ship Design conference, London, UK,* 2014.

13 ANČIĆ, I., VLADIMIR, N., RUNKO LUTTENBERGER, L., 'Energy efficiency of ro-ro passenger ships with integrated power systems', *Ocean engineering, 166, 350-357*, 2018.

14 AGENCIJA ZA OBALNI LINIJSKI POMORSKI PROMET, 'Promet putnika i vozila na državnim linijama', http://www.agencija-zolpp.hr/Portals/12/download/Promet_putnika_i_vozila_2017.pdf, 2018.

15 https://www.croatiaferries.com/island-hopping-croatia/, accessed 5.11.2018.

8. AUTHORS BIOGRAPHY

Maja Perčić was born 1993 in Pula, Croatia, where she finished grammar school in 2012. In the same year she enrolled into Chemical Engineering studies at the Faculty of Chemical Engineering and Technology, University of Zagreb, where she defended her master thesis in 2017. Since October 2018 she works as a research assistant at the Chair of Marine Engineering at the Faculty of Mechanical Engineering and Naval Architecture, University of Zagreb (UNIZAG FSB), where she started her PhD studies in the field of Naval Architecture and Ocean Engineering. Her research is funded by the Croatian Science Foundation within the GRiMM Project (http://grimm-project.com/en/home/) and her research interest include: energy efficiency of marine power systems, ship environmental eligibility, life-cycle assessments of ships and marine transportation in general, etc.

Ivica Ančić, Ph.D., a postdoc at UNIZAG FSB, obtained his PhD in 2016 in "Energy efficiency and environmental impact of integrated power systems". He is involved in several research projects funded by Interreg Adrion, European Social Fund, Croatian Science Foundation etc. His interests are in marine power systems, energy efficiency evaluation and environmental impact assessment. So far he published over 20 scientific papers, of which 8 in scientific journals.

Nikola Vladimir, Ph.D. is Assistant Professor and Head of the Chair of Marine Engineering at UNIZAG FSB. He obtained MSc degree in Naval Architecture in 2007 at UNIZAG FSB, and PhD in the field of Hydroelasticity of ship structures in 2011, at the same University. His research interest include: ship design procedures, ship energy efficiency and environmental eligibility, vibration theory, numerical methods, dynamic analysis ship structures (hydroelastic, engine & propeller induced, etc.), vibration phenomena in ship propulsion system (main engine, shafting, propeller…), as well as all other dynamic loading and response issues inherent to ships and offshore units. As a guest researcher he was appointed at the Technical University of Lisbon, Instituto Superior Tecnico, Lisbon, Portugal, Research Department of Bureau Veritas, Paris, France and at the Pusan National University, Busan, South Korea. Dr. Vladimir is a member editorial boards of several international journals, member of International Ship and Offshore Structure Congress (ISSC) as well as the member of the Scientific Council for Maritime Affairs of Croatian Academy of Sciences and Arts. He published over 100 papers in journals and specialized conference proceedings.

HOW TO REACH TOTAL POWER TRAIN EFFICIENCY

M Koli and **J Rekola,** Steerprop, Finland

SUMMARY

When making a shift to new and greener technologies, the best approach should be one that examines the broad picture – the entire drive train.

All too often, when designing a purely hybrid or hybrid-electric propulsion system, designers and suppliers spend most of their efforts focusing on the individual components of the system. They look for ways to make each of the multiple components as efficient as possible. The amount of power needed for a generator is calculated in excess.

We have mastered exceptionally high hydrodynamic efficiency for any type of vessel, utilizing permanent magnet machines to add to overall efficiency, lower operational losses and optimize the entire system.

To effectively move a vessel through the water, every component counts, but the best approach is to consider the entirety in order to achieve total power train efficiency.

1. INTRODUCTION

What happens under the water? What exactly leads to hydrodynamic propulsion efficiency? Is it the efficiency of the components when viewed individually? No. That alone does not create the advanced synergies propulsion efficiency requires for any significant sustainable breakthrough.

Steerprop's approach is to consider together all the critical components, such as the frequency converters, generators and propellers. We also look into the cooling and excitation auxiliary equipment – normally forgotten, though they play a significant role.

Steerprop's power train optimization starts at the propeller, where mechanical energy is converted to thrust, to propel the vessel. If the propeller is inefficient – or the best propulsion efficiency is not achieved for every type of vessel operation – the rest of the power train needs to be over-dimensioned in order to transmit the power required for different propulsion needs. When the power train is over-dimensioned, it is constantly operating in partial load status, and the best possible efficiency cannot be achieved throughout the power train. Figure 1 illustrates from left to right how power train losses are born. Looking at it from right to left shows how, by increasing efficiency on the right, the losses on the left automatically become smaller.

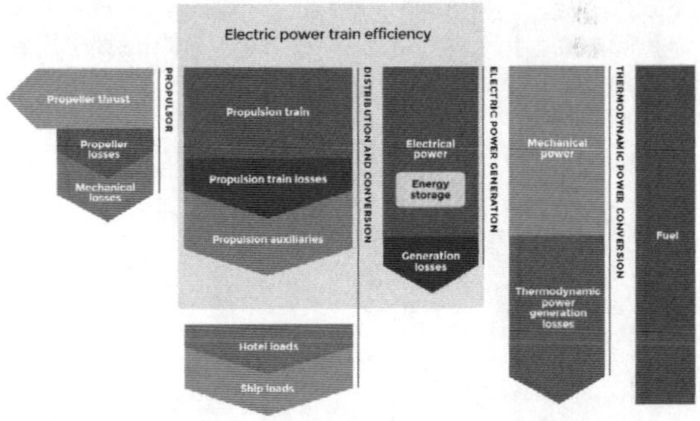

Figure 1. The basis of Steerprop's efficiency doctrine: fuel to thrust chart

We have created tools to help shipowners and design offices analyze the efficiency of their entire drive train, to reach the best possible performance at sea.

2. SCOPE OF ECOSYSTEM

Our scope of delivery in the drive train ecosystem includes the WE Tech propulsion drive solution for economical operations. Efficient power distribution is based on the LM Unit Series from Steerprop, the WE Tech Propulsion Drive™ and permanent magnet generators.

Using this solution, the vessel's electrical power is generated with permanent magnet variable-speed generators connected via inverters to a common DC link. The generators and battery bank can operate continuously in parallel.

An energy storage solution, such as a battery, can be used for peak-shaving functionality by supporting the main generators and diesel engines – taking on fast load changes (load peaks) thus keeping main generator electrical load constant. This ensures efficient, surge-free operation of the main generators, while all large consumers are provided a very fast load response. The electrical network is kept stable.

An energy storage solution has a relatively high cost per kWh, so it is important to use the energy storage solution for as many modes as possible. One of these is the spinning reserve mode, which provides an energy reserve for power generation. By using the energy storage solution as a spinning reserve, one can limit unnecessary parallel running of generators.

Alternative power sources, such as solar panels, can be connected via a DC-DC converter to the common DC link.

A ship-to-shore connection can be made via a transformer to the WE Tech Propulsion Drive™.

Bow thruster, FiFi pump and deck machinery drives can be connected to the common DC link of the WE Tech Propulsion Drive™, to allow a minimum amount of AC/DC conversions in the power train.

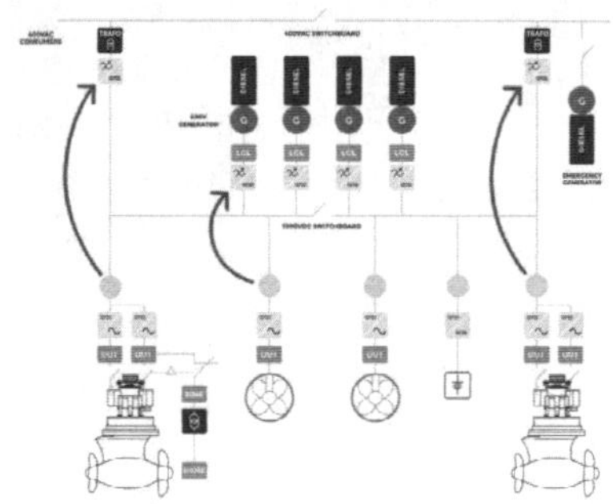

Figure 2. Description of the Steerprop Ecosystem

2.1 PROPULSION SYSTEM

The main propulsion system for the vessel is composed of two mechanical azimuth propulsors. Propulsors are of the LM type. The type and its name come from the mechanical power train configured in an L shape, and M stands for the integrated permanent magnet motor. The parts of the propulsor below the waterline have also gone through a redesign process to ensure the best possible hydrodynamic efficiency, combined with the long proven reliability of Steerprop propulsors.

Depending on the operational profile of the vessel, different propulsion configurations can lead to much better hydrodynamic performance. For example, a ducted propeller design will be beneficial if vessel operation consists of very high running hours in DP operation, or whether a high bollard pull requirement is the dominating factor in vessel specification. If high open-water efficiency for transit is required, a Steerprop CRP unit would offer the best performance in that case. CRP stands for contra-rotating propellers. The unit is equipped with two propellers, one at each end of the unit. Figure 3 shows a Steerprop CRP unit. With contra-rotating propellers, the rear propeller is able to absorb more than 90% of the front propeller's race swirl energy and turn it into forward thrust.

Figure 3. Steerprop CRP unit

A CRP unit outperforms an open single-propeller unit in bollard pull conditions and offers better performance in both open water and bollard pull operations. With a CRP unit propeller, for example, loading can be reduced if the propeller diameter cannot be increased due to draught restrictions. The CRP unit's combined propeller surface area is larger, compared to single-propeller units. Figure 4 illustrates an ideal propeller efficiency curve as a function of propeller loading per square meter. This curve shows a difference of 7.5% in propeller hydrodynamic efficiency between the CRP and the open single-propeller unit, due to the lower loading of the propeller in the CRP unit.

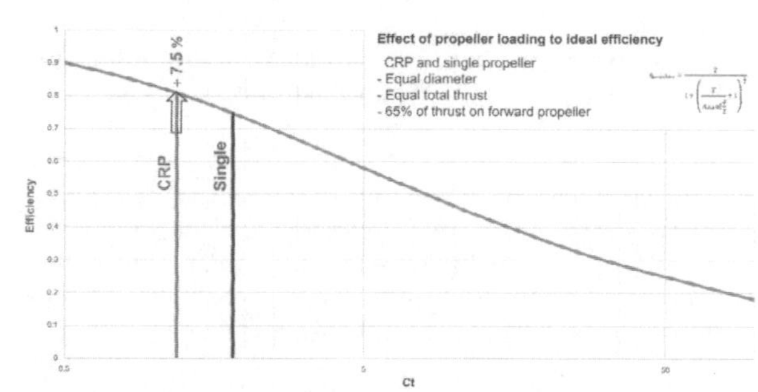

Figure 4. Ideal propeller efficiency as a function of its loading

The differences in fuel consumption between CRP and ducted units have been calculated with a fictional operational profile for a vessel with 2 times 1,100 kW main propulsion units. The results of the calculation for one week's operation are shown in Appendix 2. From the graph, it can be seen that a CRP unit's strengths are at higher vessel speeds – above 5 knots. A unit with a single ducted propeller has higher performance during maneuvering and slow speed operations due to better attributes for bollard pull conditions.

Regardless of the propulsion unit propeller configuration, every unit can be equipped with a high-efficiency permanent magnet motor. This feature is available because of Steerprop's modular propulsor design. All propulsors can be built as traditional Z-drive propulsors, with two right-angle gears, or as an LM unit with only the lower right-angle gear. With the LM unit, the propulsion unit does not need the upper gear required in a traditional Z-drive azimuth propulsor. With deletion of the upper gear, the mechanical power train efficiency is increased by 1–1.5%.

Permanent magnet motors integrated with the propulsion unit are always designed to best fit each specific propulsion unit. This enables the highest possible efficiency compared to standard induction motors, normally used with Z-drive azimuth propulsors. With the LM units especially, the motor's partial load efficiency will improve significantly. Appendix 2 illustrates the electrical efficiency of a traditional induction motor and a permanent magnet motor as a function of vessel speed. The required propulsion power is shown in grey in the graph. With permanent magnet motors and a vessel speed at 60% of the design speed, the gain in electric motor efficiency is almost 10%. This illustrates the gains from a permanent magnet motor when operating on partial loads.

Using Steerprop LM units with integrated permanent magnet motors brings significant weight reductions, compared to traditional Z-drive azimuth propulsor and induction propulsion motors. The weight difference between these can be 10%, excluding the induction motor foundation, which will increase the difference in the LM unit's favor.

2.2 PROPULSION MOTOR DRIVE

To achieve overall efficiency for the whole propulsion drive train, the permanent magnet motor and variable frequency drive configuration needs to be designed so that the combined losses are as low as possible.

After the propulsion motor and the inverter supplying it have been matched for the best combined efficiency, power train optimization continues to the rectifier unit. For a long time, the industry standard has been for a passive rectifier comprised of a diode bridge and possible choke filter. A passive rectifier is a very cost-effective unit but does not enable the best possible performance. Figure 4 illustrates diesel-electric propulsion powers with AC grid and main switch board. All major consumers have a separate AC supply from the main switch board and separate transformer, rectifier and inverter power lines before the motor.

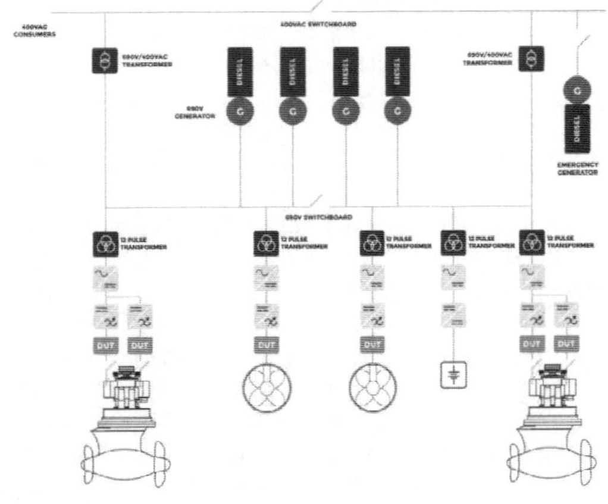

Figure 5. Diesel electric propulsion power train with AC grid and passive rectifiers

For higher efficiency and controllability of the rectifying process, it is better to use an active rectifier – or as it is commonly referred to, an active front end (AFE) rectifier. Usually, the biggest contributor to the use of AFE is to limit total harmonic distortion (THD) in the supplying grid. High levels of THD cause higher losses in the transfer grid, generators and in other consumers. Figure 5 shows the power train with AFEs substituting for the passive rectifier and 12-pulse transformer.

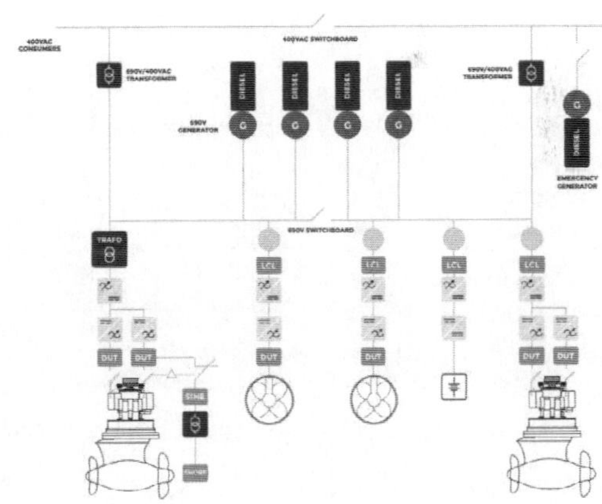

Figure 6. Diesel electric propulsion power train with AC grid and active front end rectifiers

The heavy propulsion transformer is not needed with AFE, so weight savings can be gained through AFE technology, contributing to either higher net tonnage of the vessel or smaller gross tonnage.

The second benefit of using AFE instead of a passive rectifier is the possibility for regenerative braking, or in other words, two-way power transfer. With AFE technology, for example, the propulsion converter can be utilized as a shore connection since propulsion power is not needed. This solution is shown in figures 4–6. When the shore connection is made via the converter, the shore-side frequency and voltage do not need to be the same as on board. Therefore, changing from or to a shore supply can be done with such precise control that it almost excludes the possibility of a blackout during the process.

2.3 VARIABLE-SPEED POWER GENERATION

When generators are directly connected to the main switchboard of the vessel, the switchboard is naturally an AC type. This solution then dictates that the generators' electric output frequency be of a certain constant, normally 50 or 60 hertz, and will also fix the diesel engine rotational speed to the corresponding constant. A vessel's grids – other variables, voltage and power factors – are controlled with the generator's magnetization to keep the grid in balance. Transferring the AFEs from the main consumers to be connected directly to the generators, the rotational speed of each generator can differ between them. Figure 6 shows the changes in the distribution system. Each generator has its own dedicated AFE, and the vessel's auxiliary electric grid will be established with two AFEs equipped with certain features, so they become grid converters.

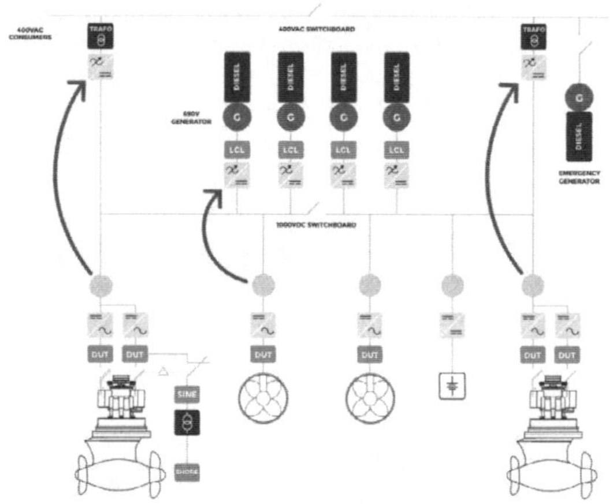

Figure 7. Diesel electric propulsion power train with DC grid

With individual AFEs for each generator, the generator frequency, as already stated, does not have to be same as in others. But this also applies to voltage and power factors, since the generators' AC systems are not any more interconnected. This transformation will lead to DC grid in the main electrical distribution system, in which there is a single main variable to control – voltage.

Since generators do not have to control the voltage nor the power factor in the main grid, the generator does not need controllable magnetization. With this fact we are also able to use permanent magnet electrical machines in electricity production, as well as on the propulsion motor side. Permanent magnet generators lead to better efficiency in the rotational energy transformation to electricity, especially in the partial loading of generators. Partial loading of diesel generators does not lead to the best possible fuel efficiency, i.e. specific fuel consumption. The situation improves when the rotational speed of the motor can be decreased, and therefore its relative loading will increase and specific fuel consumption will reduce.

Power electronic devices have the same feature as almost all machines: the efficiency of the device decreases if it is used in partial loading – in other words, not in the conditions for which it was designed to operate. The rectifiers will operate closer to the optimal point of loading when working as generator-specific devices. Generator loading is kept at the best possible figure at all times to keep the diesel specific fuel consumption low. In slow steaming, for example, propulsion power can be 30% of nominal power, and generator loading can still be kept at 80% by operating only the needed number of diesel gensets. This allows the AFE to operate at much better loading and efficiency, compared to propulsion-line specific rectifiers.

PERMANENT MAGNET GENERATOR

Advantages of the permanent magnet motor, compared to externally excited synchronous motors, come from the structure of the rotor. In both generator types the stator winding and structure are made with the same design, but the rotor has significant differences. With external excitation, the rotor has its own coils to produce the magnetic flux in the motor's air gap. In a permanent magnet machine, the magnetic flux is created with the magnets, as the name states, hence no separate coils are needed. This simplifies rotor design greatly and makes it lighter. It also minimizes losses in the rotor, since no losses are generated due to magnetization current in the rotor windings. With a simple rotating part design, the maintenance of the permanent magnet generator is easy, since wear parts are minimal, mainly the bearings.

2.5 DC-DC CONVERTER FOR ESS

The energy storage system (ESS) is installed to enable the highest possible power generation efficiency by allowing diesel generator loading control without any change in the consumption of electricity. The number of gensets used can be adjusted so that power production on board is sometimes higher than consumption and sometimes lower than consumption. During these periods, the optimum number of gensets is operated to achieve the best efficiency in transforming the chemical energy of oil into electricity. To ensure the least amount of energy loss in this process, the ESS must operate at high efficiency. Also the link between a vessel's grid and its ESS needs to be effective.

When operating with a traditional AC grid filter, a rectifier and DC-DC converter are needed to connect the

ESS into the power system. With DC distribution only, a DC-DC converter between the main grid and ESS is needed. This ensures better efficiency for the coupling of the ESS to the grid and allows additional energy savings both ways: when charging and discharging the battery. If efficiency is important in the charging operation, it is even more important during discharging. This is due to the fact that all the lost energy in this part of the process is virtually reducing the ESS capacity, because part of the stored energy is turned into heat at the coupling.

2.6 EFFICIENT AUXILIARIES FOR THE PROPULSION POWER TRAIN

Auxiliaries needed in the whole propulsion power train are limited to the mechanical azimuthing unit's lubrication and cooling, and to electric power train cooling. Lubrication and cooling for the propulsor's mechanical transmission is accomplished with variable-speed-driven lubrication pumps. A pump's speed is controlled according to the current needs for cooling or pressure. The electrical propulsion power train's equipment is all liquid-cooled to minimize the need for cooling air fans and the use of power for cooling. The need for cooling is lower when the whole power train has been designed to maximize efficiency and save energy.

Azimuthing propulsors steer the vessel, and their turning gear is electromechanical. Electric motors turn the propulsor. The motors are driven by variable frequency drives. In this solution, only the power needed for each specific operation is used. For example, during autopilot operation in calm weather, only 3% of the turning gears power is needed. Compare this to hydraulic steering gear for an azimuthing propulsor: 30% of the nominal power. The difference in power requirements is significant

2.7 AUXILIARY DEVICE DRIVES

Due to the high power of the bow thruster, it is best to feed it from the DC grid with a low harmonic drive. If fed from an AC grid with a 6-pulse frequency converter, the result is high THD levels. For other consumers, which would in normal cases be fed by frequency converters, the benefit pf using the DC grid and low harmonic drives is a low THD contribution. Also, if a variable frequency drive feeds the bow thruster, you can use a fixed-pitch propeller instead of a controllable-pitch propeller, which is beneficial due to lower capital expense and less maintenance.

When connecting, for example, deck machinery to a DC grid via variable frequency drive, this offers the option of regenerative braking back to the grid. With this feature it is possible to load the battery when releasing the winch with its load, and then reuse the energy when the winch line needs to be reeled in.

2.8 GRID CONVERTER

A grid converter establishes the AC grid for the vessel's other services, which operate with alternating current. With the grid converter, variable-speed power generation is also available for the hotel load and ensures better fuel economy. In addition, it allows the vessel's main power supply to be the DC grid. Therefore, the THD that would be created from bigger consumer's variable frequency drives can be isolated from the auxiliary power grid. This increases the efficiency of the hotel load grid and consumers and reduces noise in the grid for sensitive appliances.

2.9 CONVERTERS FOR ALTERNATIVE POWER SOURCES

Many alternative power sources generate DC current as their natural source, such as solar panels or hydrogen cells. With a DC grid solution, these sources can be connected to the grid with only DC level adjustment and minimal losses.

Also, other alternative power sources, like exhaust gas or miniature steam turbines together with a generator, can be connected to the DC grid with the AFE – enabling variable speed operation for the turbine, and therefore enlarging the turbine's efficient operational area.

3. CONCLUSIONS

Diesel-electric propulsion, together with modern semiconductor devices and permanent magnet motors, provide much greater flexibility for the whole propulsion power train design. When combining this with careful design and thoughtful decision on alternatives for the propulsion source, the efficiency of the whole propulsion line can improve significantly.

A major factor in the overall design of the power train is quite the same as for the whole vessel. Therefore, the main question should be – what is the intended use of the vessel? Will it be steaming steadily between ports; will it be held in dynamic positioning mode as a standby vessel, or something between those two? This will greatly change the focus point. If operations are performed mostly at about the design speed of the vessel, power should be maximized for the best possible efficiency around that point, since not much will be gained by partial load optimization due to low running hours.

Correct decision and dimensioning of the propulsion unit and power is the key factor for creating a power train with the features needed for excellent operational readiness with the best possible efficiency. If the power is not sufficient, vessel design speed might not be met, or sea-keeping properties could be too weak. But if propulsion power is calculated with margins that are too large, that will lead to continuous operation with significant partial loading and decreased efficiency throughout the power train.

A DC grid solution together with the correct propulsion configuration offers the flexibility for any vessel to be more efficient in a wider range of different operational profiles. This will benefit vessel operators during the life cycle of the vessel, since it can be deployed to different jobs without compromising fuel efficiency and increasing costs. An optimized solution makes it possible to use diesel engines in a more flexible manner and control the power input from different sources easily. In addition, it allows the use of alternative sources of power to reduce the carbon footprint of vessels, therefore meeting the ever-increasing CO_2 emission regulations.

4. AUTHORS BIOGRAPHY

Mika Koli, Sales Director, Steerprop Ltd.

Mika has devoted his 20-year career to help marine and industrial customers build energy efficiency in power production and consumption. To solve their challenges in these matters, he has been a strong advocate for choosing green solutions that add sustainability over the long run.

Since March 2018, Mika has been Sales Director at Steerprop, one of the world's leaders in azimuth propulsors. Prior to this position, he was Business Development Manager, Marine at Yaskawa Environmental Energy / The Switch, where he was responsible for developing the marine business and for applying the company's permanent magnet drive train technology to advanced marine solutions.

Earlier, Mika worked at various companies in the marine industry, including engineering offices, equipment manufacturers and system integrators in Finland and Germany. Mika graduated with a BSc in machine automation and industrial economics from Turku, Finland, and completed his MBA at Swansea in the UK.

Juho Rekola, Chief Design Engineer, Steerprop Ltd.,

Juho has been involved in marine technology from the beginning of his career. His focus has always been on high-efficiency solutions for the industry. Previously, Juho worked as a Chief Design Engineer, Electrical, at Alfa Laval Aalborg for its cruise unit's key customer team.

Today, Juho is working with Steerprop on the electrical design of the company's LM units and hybrid power train solutions. He works closely with different stakeholders in electrical and hybrid power to give the marine industry the best possible solutions for tackling climate change and rising operational costs.

Juho has graduated with an MSc (technology) in power electronics from Tampere University of Technology, faculty of Electrical Engineering.

10. Appendix

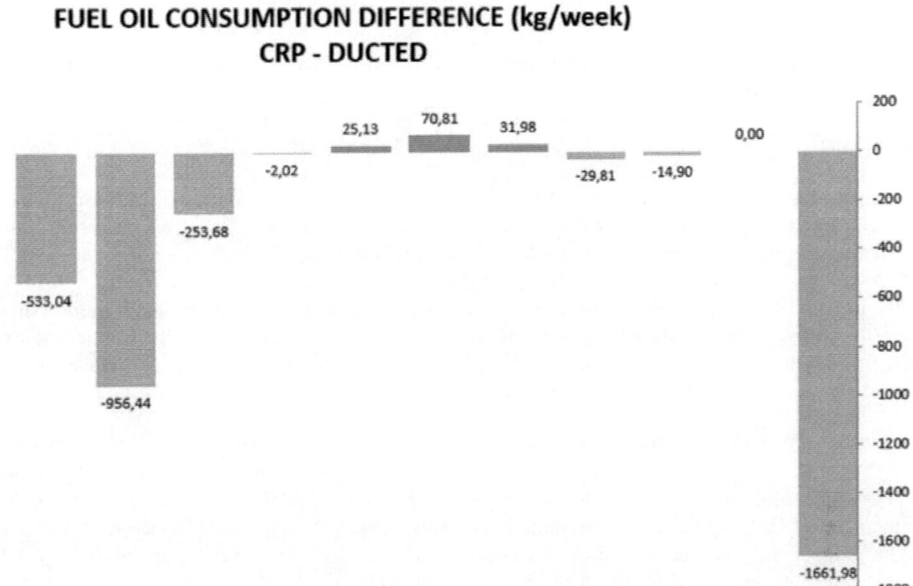

	Sailing	Sailing	Sailing	Sailing	Manouvering	Manouvering	Manouvering	Fishing/towing 10t	Fishing/towing 10t	In port / Waiting for weather	SUM
Speed (knots)	13	11	8	4	2	2	2	4	4	0	
CONSUMPTION CRP-DUCTED	-533,04	-956,44	-253,68	-2,02	25,13	70,81	31,98	-29,81	-14,90	0,00	-1661,98

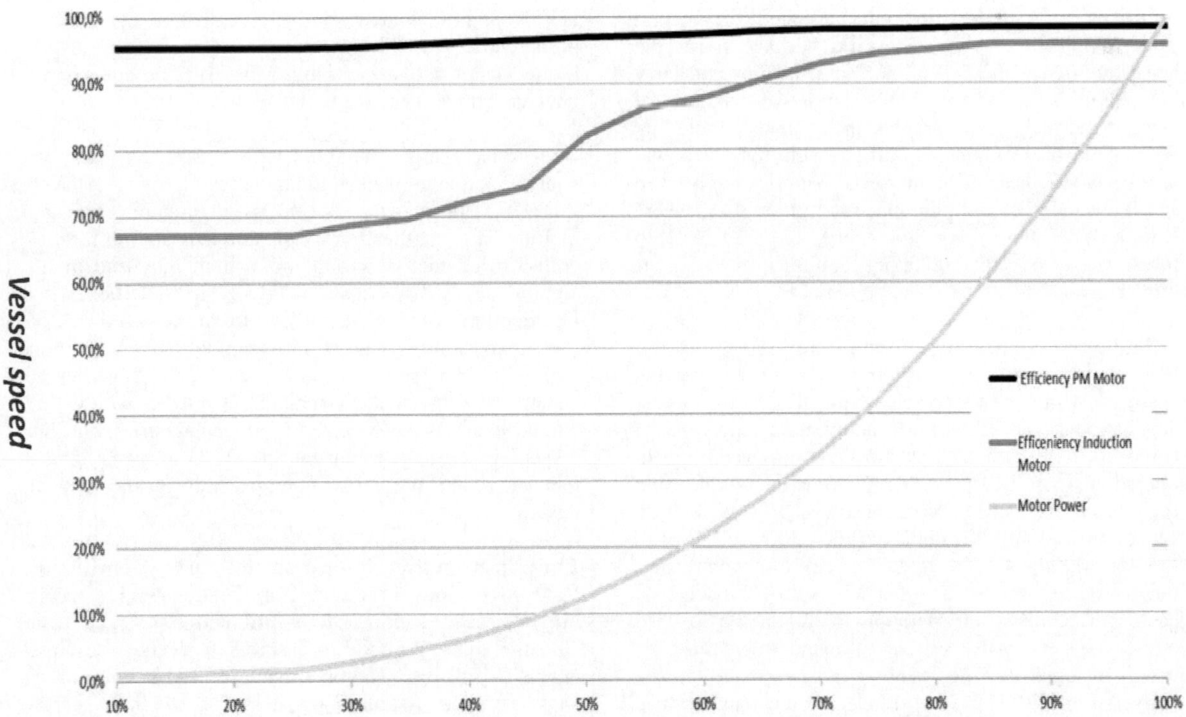

Comparison of diesel-electric with hybrid-electric propulsion system safety using System-Theoretic Process Analysis

V Bolbot, G Theotokatos, E Boulougouris, D Vassalos, Maritime Safety Research Centre, The University of Strathclyde, UK

SUMMARY

Cruise ship industry is rapidly developing, with both the vessels size and number constantly growing up, which renders ensuring passengers, crew and ship safety a paramount necessity. Collision, grounding and fire are among the most frequent accidents on cruise ships with high consequences. In this study, a hazard analysis of diesel-electric and hybrid-electric propulsion system is undertaken using System-Theoretic Process Analysis (STPA). The results demonstrate significant increase in potential hazardous scenarios due to failures in automation and control systems, leading to fire and a higher number of scenarios leading to propulsion and power loss in hybrid-electric propulsion systems than on a conventional cruise-ship propulsion system. Results also demonstrate that STPA enhancement is required to compare the risk of two propulsion systems.

1. INTRODUCTION

Developments over the recent past have driven the maritime industry towards reducing exhaust gas emissions and fuel consumption. Specific areas have been designated, the so-called Emission Control Areas (ECAs), where stringent limits for NOx and SOx emissions are applied [1]. At the same time, considerable reduction in the attained Energy Efficiency Design Index (EEDI), which is used to depict the vessel CO_2 emissions, is required by the International Maritime Organisation (IMO) from new built vessels [1]. In addition, the maritime industry is going through periods of high fuel prices, resulting in high operating costs. Furthermore, in 2018, Norway has adopted a resolution to achieve zero-emissions in world heritage fjords the latest by 2026 with application to cruise ships and ferry vessels [2]. The above render attractive the use of alternative fuels and propulsion systems, including Hybrid-Electric Propulsion (HEP) with hybrid power supply, where diesel-generators and batteries are used to ship power needs, and pure electrical propulsion, where batteries are used to store the energy required for ship functions, to meet the regulatory requirements in a cost-effective way.

Hybrid-electric and pure electric propulsion systems have already been applied on a number of existing vessels, while new vessels with HEP are under development. *MV Viking Lady*, an offshore supply vessel equipped with 500 kWh battery system has been in operation since 2013 [3]. *MV Ampere* is the world's first fully electric battery powered ferry vessel with battery capacity of 1,040 kWh deployed on a route in Norway [4]. *MV Hallaig*, *MV Lochinvar* and *MV Catriona* ferries, three sister vessels with 700 kWh battery capacity on each, are currently in operation in Scotland [5]. Two cruise ships with HEP system, allowing ship sailing by using batteries for 30 minutes are expected to be delivered in 2019 [6]. A hybrid-electric icebreaker cruise ship is under development by *PONANT*, *Stirling Design International*, *Aker Arctic* and *VARD* with scheduled ship delivery in 2021 [7]. Considering that battery technology is being constantly developed, with increasing energy density and decreasing procurement cost [8], it can be expected that use of batteries will be extended to larger size cruise ships in the near future.

The HEP achieves energy efficiency improvement by running D/G sets at optimum load by peak load shaving and functioning as spinning reserve [8-10]. Implementation of HEP leads to D/G sets downsizing, which also supports D/G sets operation at their most efficient load ranges [8]. Other advantages include higher redundancy in system and lower emissions due to charging of batteries from local grid in harbour [8, 10]. Disadvantages include relatively high cost of batteries procurement [8, 10], large batteries size and weight [9], limited number of recharging cycles [9] and addition of new hazardous scenarios to the system [8].

On cruise ships though, with passenger number equivalent to a number of inhabitants of a town, ensuring safety of propulsion system is paramount as any malfunctions may lead to propulsion loss and, in turn, to collision, contact or grounding, which may end up in significant human loss [11-13]. In addition, the introduction of batteries can lead to an increased risk of fire, explosion and crew intoxication [8]. A fire on hybrid-electric tugboat occurred due to malfunction of Battery Management System[14], whilst a number of similar events have occurred in other industries. In this respect, it is crucial to ensure that all these scenarios are identified and properly addressed during the system design.

The primary reference for designing safe systems is the IMO regulations [15] and classification society rules [16]. However, additional hazard and risk assessment studies may be required to ensure safe design and class approval [15, 16]. The only available and known safety study on HEP system is given in [3], which is a high level study. Other studies have referred to potential safety issues on HEP systems but did not follow a hazard identification method for their analysis [14]. Pertinent literature reveals the research gap, which is a hazard analysis of HEP system using well-established or novel methods and comparison with the standard diesel-electric propulsion. The research gap leads to the aim of this study, which is to analyse the safety of HEP system using System-Theoretic Process Analysis and to compare it with standard Diesel-Electric Propulsion (DEP) in terms of the developed hazards, number of potential hazardous scenarios and causal factors.

This paper is organised as following. In section two, the selected method and the rationale behind the method are presented. In section three, a short description of the system and system functionalities is provided. In section four, the analysis results and safety recommendations are given. In section five, the main findings of this study are summarised.

2. METHODOLOGY

Hazard identification and analysis is the process of defining all possible scenarios or sequences of events, which can lead to a hazard realisation [17]. A number of traditional methods can be used for analysis of power propulsion systems including Preliminary Hazard Analysis (PHA), HAZard and Operability studies (HAZOP) and Failure Modes and Effects Analysis (FMEA) [17]. However, these methods have been criticised for not addressing properly the automation functions in the system [17-20]. Control and automation though has an important role for power generation on cruise ships using either standard or diesel-electric or hybrid-electric propulsion system [21]. For this reason, the System-Theoretic Process Analysis (STPA) method has been selected for hazard identification. Another advantage of STPA is that it can be implemented on functional level, not requiring the exact details of the system and vessel. In this way the identified hazardous scenarios will have applicability to other ship types and similar propulsion systems. The method steps are presented in Figure 1 and described in more detail in the following.

STPA defines the accident as: "an undesired and unplanned event that results in loss, including loss of

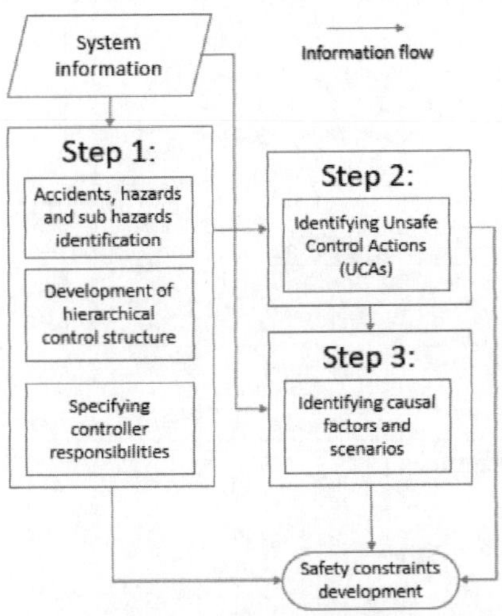

Figure 1 STPA steps.

human life or human injury, property damage, environmental pollution, mission loss, financial loss, etc." [22]. The hazards in the STPA framework are understood as: "system states or set of conditions that together with a worst-case set of environmental conditions, will lead to an accident" [22]. The hazards in STPA are viewed on a system level, so they go beyond the single failures that may occur in the system and should be referred to a specific state of the system. Sub hazards are considered states in a worst-case scenario leading to hazard realisation. Generic requirements can be specified, based on the hazards and sub hazards.

The development of a functional control structure is one of the differentiating points of the STPA analysis, compared with the other methods [22]. Usually, it starts with a high-level abstraction of the system and proceeds to a more detailed system description. The initial control structure consists of the high-level controller, the human operator and the controlled process with the basic control, feedback and communication links. A more detailed description would incorporate a hierarchy of controllers. Both high-level and detailed control structure can be used for the safety analysis at different system design stages.

After the development of the basic control structure, the next step is its refinement. The required actions include the identification a) of each controller responsibilities; b) of the process model with process variables and potential process variable values; c) of the control actions; d) the behaviour of the actuators; e) the information from the sensors; f) the information from the other controllers.

The actual hazards identification starts by finding the Unsafe Control Actions (UCAs). The possible ways to proceed are either by using the control actions types as initially proposed for the STPA [23] or by using the context tables as proposed in [18]. Herein, the second of the two approaches has been selected. According to both approaches, the possible UCAs can be of the following seven types [22]:

- Not providing the action leads to a hazard.
- Providing of a UCA that leads to a hazard.
- Providing the control action too late.
- Providing the control action too early.
- Providing the control action out of sequence.
- Control action is stopped too soon
- Control action is applied for too long.

According to the STPA, there is also another type of UCA, when the safe control action is provided but is not followed. This type of failure mode is addressed during the identification of causal factors in the second step of the method. Similarly, with the system hazards, safety constraints can be derived for the UCAs, aiding the identification of possible safety barriers.

The second step in the hazard identification of the STPA has the purpose of determining all the scenarios and causal factors leading to the UCAs. This is done by examining the hazardous scenarios including software and physical failures as well as design errors. There are several ways to organize the results of the hazardous scenarios by using tables or lists. In this work, the process was augmented by a checklist, developed on the basis of previous studies [24, 25]. The main categories of causal factors are:

- Inappropriate control input
- Hardware failure
- Software faulty implementation
- Software faulty design
- Erroneous or missing input
- Inadequate control command transmission
- Flawed execution due to faults in actuator or physical process
- Conflicting control actions

3. SYSTEMS DESCRIPTION

The conventional diesel-electric and hybrid-electric propulsion system single line diagram is presented in Figure 2 whilst functional control structure for both systems is given in Figure 3. Two switchboards and engine rooms are available to comply with Safe Return to Port rules requirements [26]. The power network is of the Alternate Current type. It has been also assumed that DEP plant operates with the bus-tie circuit breaker connected. Power Management System (PMS) starts/stops the engines based on ship consumers electric load demand. Switchover between the plant Diesel Generators (D/G) is implemented based on the D/G sets running hours. The PMS can implement a fast-electrical load reduction for the propulsion motors and bow thrusters as well as preferential tripping functions (fast

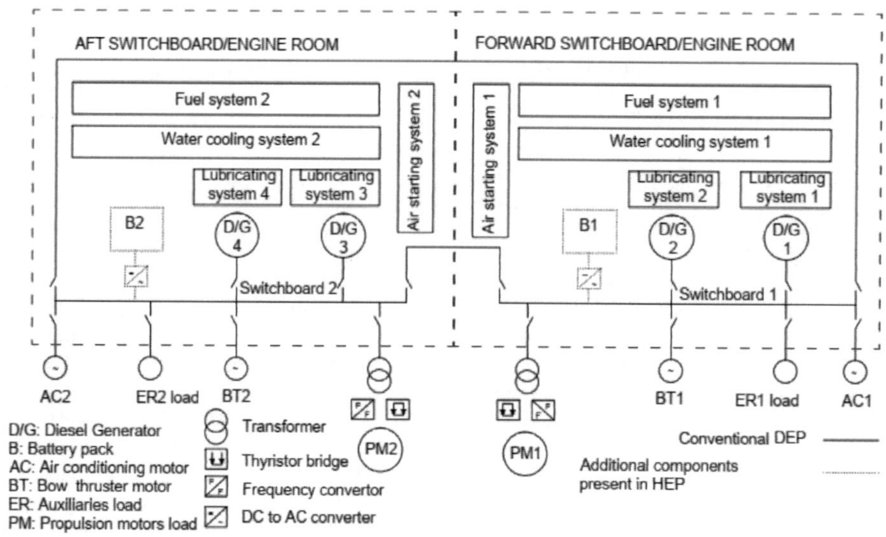

Figure 3 Single line diagram of conventional and hybrid power network.

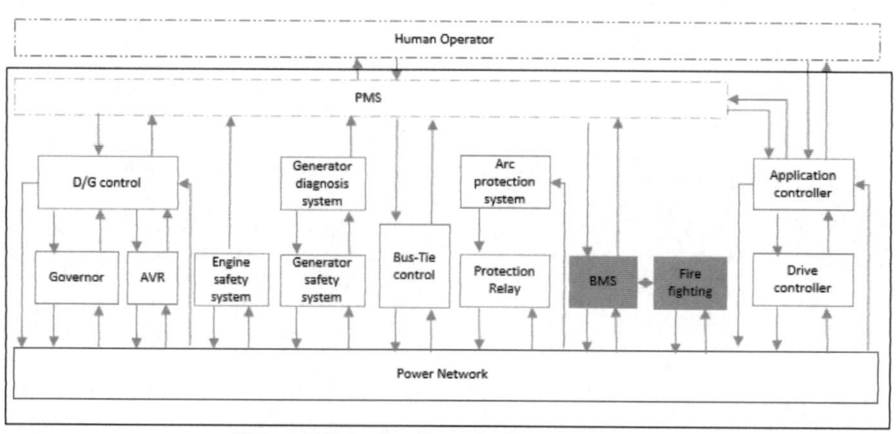

Figure 2 Functional control structure.

load reduction) by tripping Heat Ventilation Air Conditioning (HVAC) units.

The D/G sets operate in the droop mode and their power output is regulated by speed governor and Automatic Voltage Regulator (AVR). Several safety systems are used to trip D/G sets and propulsion motors if a fault had been observed. The DEP control network is also considered to be isolated from other networks, so no hazardous scenarios are developed in the system because of cyber-attacks. It is also considered that the human operator neither reduces nor introduces new hazards.

In the investigated HEP system, in addition to the components present in conventional DEP, one battery pack per switchboard with current converter is considered. The battery output and condition are controlled by a dedicated Battery Management System (BMS) which monitors the actual battery health state and the battery and cell capacity and controls the battery cells charge status, the discharging/charging rate, the converters power output and the battery auxiliary systems. The BMS communicates with PMS to determine the actual power status and power demand implementing in this way the Energy Management System functions. The BMS also communicates with fire-fighting systems to determine the battery operating status. Battery capacity is considered adequate to cover the whole ship power demand for a limited period. The battery has been considered of Li-Ion type.

4. RESULTS AND DISCUSSION

Based on previous Formal Safety Assessment studies, the following causality scenarios can be considered as accidents [27]:

- Collision [A-1]
- Contact [A-2]
- Grounding [A-3]
- Fire [A-4]
- Explosion [A-5]
- Machinery damage [A-6]
- Foundering [A-7]
- Operating personnel injury or death [A-8]

These accidents are not fully disjoint, as a fire can lead to collision and vice versa [28]. In addition, numerous hazards can be connected to the accidents on a cruise ship and there can be interactions between different hazards. Herein, the most important and those related to the system under analysis are referred to [11, 27]:

- Propulsion loss [H-1] leading to collision, contact and grounding accidents. The propulsion loss can be further developed into the following sub hazards:
  - D/G sets overload [H-1-1].
  - Transients [H-1-2].
  - Imbalanced power generation [H-1-3]
  - D/G sets unavailability [H-1-4]
  - Batteries unavailability [H-1-5]
  - Propulsion motors unavailability [H-1-6]
- Flammable liquid on hot surfaces in the engine room and other conditions leading to [H-2] fire in engine room.
- Uncontrolled electrical faults in equipment leading to [H-3] fire and explosions in system components or blackout (propulsion loss).
- Toxic/flammable atmosphere in battery room leading to crew intoxication or fire [H-4].
- Anomalous conditions in batteries leading to fire and thermal runaway [H-5].
- Arson – deliberate act resulting in fire [H-6].
- Human erroneous operation [H-7]
- Cyber-attack leading to any of previous hazards [H-8].
- Water ingress [H-9]

Although, it is acknowledged that there is contribution from hazards [H-6]-[H-9] to the overall system risk, these hazards can be considered as external to the system presented in Figure 2 and Figure 3 and thus their analysis has been omitted.

The developed control structure has been already provided in Figure 3. The difference between the two propulsion systems can be found in the presence of Battery Management System and additional interactions between the fire-fighting system and the propulsion system. The description of responsibilities of each controller and their control actions, although necessary for the analysis have been omitted for brevity purposes.

In total, 160 and 228 potential UCAs have been identified in DEP and HEP system, respectively. The increase in UCAs number can be attributed to the increase in the number of control actions implemented by BMS. However, as it can be viewed, it leads to a significant increase in the number of potential UCAs (more than 40%). The distribution of UCAs per hazards is given in Figure 4. As it can be seen from this figure, the number of hazardous scenarios other than propulsion loss, leading to fire or crew intoxication is significantly higher in the investigated HEP system in comparison with the DEP system. The number of scenarios leading to propulsion loss is also significantly higher. These results do not necessary imply that the risk level is higher in the investigated hybrid-electric system than the risk in conventional DEP system, rather that there are much more paths to the accident in hybrid-electric systems than in DEP, which must be carefully controlled. This also indicates that a successful cyber-attack on a hybrid-electric vessel will lead to more hazardous scenarios than in conventional cruise ship vessels.

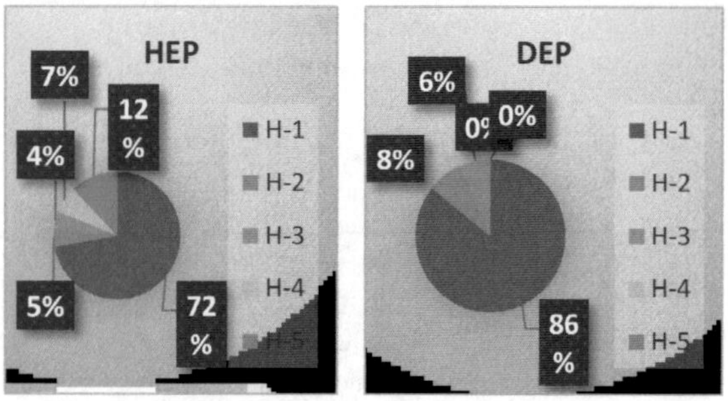

Figure 5 UCAs distribution per hazards.

The distribution of UCAs failure modes for the investigated HEP system is given in Figure 5. The results for the DEP are similar to HEP system. As it can be observed from Figure 5, the primary failure modes are related either to failure to implement the intended control action or to implement the intended action in time or commission errors. These types of control actions as well as actions applied in wrong order are related to the designed safety functions or automated control actions in the investigated system. Stopped too late, stopped too soon and applied too early were related mostly to control actions implemented by the investigated system PID controllers.

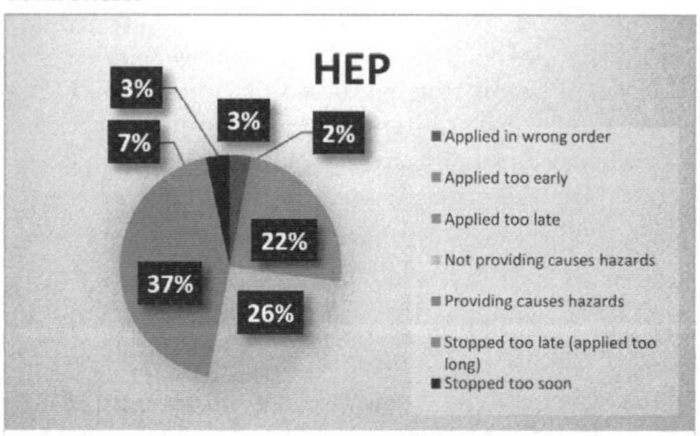

Figure 6 Failure modes distribution.

In total, 2,225 and 1,523 causal factors have been identified for the hybrid-electric and the conventional diesel-electric propulsion systems. The causal factors distribution for the HEP system is given in Figure 6. The results for the conventional DEP are similar. As it can be observed, most scenarios are dependent on installed control software errors, either controller design or implementation errors. Errors in sensors have also been identified as potential causal factors.

Based on the conducted analysis and the derived results comparison, the following safety recommendations can be made:

- In HEP systems, adequate means must be provided to prevent and mitigate scenarios leading to fire to ensure that the hazardous scenarios leading to fire do not lead to higher

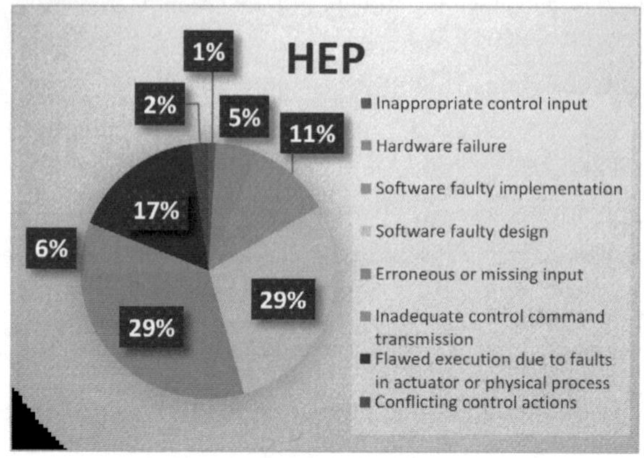

Figure 4 Distribution of causal factors.

  risk in hybrid-electric system than in the conventional system. This includes systems responsible for batteries temperature management and fire-fighting and proper selection of location for batteries.
- System operational conditions must be thoroughly understood and addressed in system design. This includes the batteries and actuators degradation mechanisms, potential modifications in the systems and software updates during operation and maintenance.
- Rigorous testing of the system control actions must be implemented for ensuring their functionality during the design, development and trial phase according to hazardous identification process in both diesel-electric and hybrid-electric system. The development and test cost in hybrid-electric system will be higher since the number of scenarios to be addressed is also higher.

From the STPA application to the conventional diesel-electric and HEP system an extensive list of safety requirements for the employed control systems has been derived. However, some of the STPA restrictions have been revealed during the analysis. Potentially, more scenarios could be identified if more refined system representation was used. In addition, STPA is applied for scenarios development, but did not allow risk estimation and scenarios ranking, so the only discussion about potential safety implications can be in terms of hazardous scenarios.

5. CONCLUSIONS

In this study, the STPA has been applied for hazard identification and analysis of a diesel-electric and hybrid-electric propulsion systems. Through its application, hazardous scenarios in automation and control system have been identified and compared.

The main findings can be summarised as follows:

- Hazardous scenarios leading to fire accidents are significantly more in HEP systems, thus they must be carefully controlled.
- Scenarios number leading to propulsion loss and potential collision, contact, grounding is also higher in HEP systems than in conventional DEP systems.
- Failure modes and potential causal factors distributions are similar in hybrid-electric and diesel-electric propulsion systems.
- Special attention must be paid to software design, software testing, sensors redundancy and batteries location in hybrid propulsion system.
- An improvement in STPA method must be considered to allow ranking of different scenarios and estimating risk.

In summary, the results indicate the high importance of proper operation of control and automation systems for diesel-electric and hybrid-electric systems safety. A potential future work could investigate the risk level in more detail by improving the STPA method.

6. ACKNOWLEDGMENTS

The work presented in this paper was partially supported by the "NEXUS – Towards Game-changer Service Operation Vessels for Offshore Windfarms" project that was funded from the European Union's Horizon 2020 research and innovation action under grant agreement N° 774519. The authors are grateful to Dr Romanas Puisa from Maritime Safety Research Centre to Dr George Psarros, Dr Ole Christian Astrup, Dr Rainer Hamann, Dr Pierre C Sames from DNV GL AS and Kevin Douglas from Royal Caribbean for their valuable comments and support. The opinions expressed herein are those of the authors and should not be construed to reflect the views of European Commission or the acknowledged individuals and their associated organisations.

7. REFERENCES

1. International Maritime Organization. Regulations for the prevention of air pollution from ships and NOx technical code 2008. Organization IM, editor. United Kingdom, London: IMO publishing; 2009.
2. GREENPORT. Norway adopts zero-emissions regulations in world heritgate fjords 2018 [Available from: https://www.greenport.com/news101/Regulation-and-Policy/norway-adopts-zero-emissions-regulations-in-world-heritage-fjords.
3. Jeong B, Oguz E, Wang H, Zhou P. Multi-criteria decision-making for marine propulsion: Hybrid, diesel electric and diesel mechanical systems from cost-environment-risk perspectives. Applied Energy. 2018;230:1065-81.
4. Corvus-Energy. World's first all-electric car ferry 2016 [Available from: https://corvusenergy.com/marine-project/mf-ampere-ferry/.
5. Ltd CMA. History of our hybrid ferries 2013 [Available from: http://www.cmassets.co.uk/project/hybrid-ferries-project/.
6. Hurtigruten. Hurtigruten names hybrid explorer ships 2018 [Available from: https://www.hurtigruten.co.uk/about-us/news/new-hybrid-explorer-ships/.
7. Dhanvijay N. Vard to build hybrid LNG cruise vessel for Ponant: Electrans; 2017 [Available from: https://www.electrans.co.uk/vard-to-build-hybrid-lng-cruise-vessel-for-ponant/.
8. Brandsaeter A, Valoen LO, Mollestad E, Haugom GP. In focus – the future is hybrid. DNV GL. 2015.
9. Räsänen J-E. Current and future scale limitation for alternative marine power and propulsion solutions. Power & Propulsion Alternatives for Ships; Rotterdam, Netherlands: The Royal Institution of Naval Architects; 2017.
10. Geertsma RD, Negenborn RR, Visser K, Hopman JJ. Design and control of hybrid power and propulsion systems for smart ships: A review of developments. Applied Energy. 2017;194:30-54.
11. Bolbot V, Theotokatos G, Vassalos D. Using system-theoretic process analysis and event tree analysis for creation of a fault tree of blackout in the Diesel-Electric Propulsion system of a cruise ship. Marine Design XIII, Volume 2: CRC Press; 2018. p. 691-9.
12. Nilsen OV. FSA for Cruise Ships - Task 4.1.1 - Hazid identification. 2005.
13. MAIB. Report on the investigation of the catastrophic failure of a capacitor in the aft harmonic filter room on board RMS Queen Mary 2 while approaching Barcelona 23 September 2010. United Kingdom, Southampton; 2011.
14. Hill DM, Agarwal A, Gully B. A review of engineering and safety considerations for hybrid power (Lithium-Ion) systems in offshore applications. Oil and Gas facilities. 2015;June 2015:68-77.
15. International Maritime Organization. SOLAS: consolidated text of the International Convention of Safety of Life at Sea, 1974, as amended. 6th consolidated edition ed: International Maritime Organization; 2014. 420 p.
16. DNV GL. Additional class notations: Battery power-Part 6 Chapter 2 Section 1 2018.
17. Bolbot V, Theotokatos G, Bujorianu LM, Boulougouris E, Vassalos D. Vulnerabilities and

safety assurance methods in Cyber-Physical Systems: A comprehensive review. Reliability Engineering & System Safety. 2019;182:179-93.
18. Thomas J. Extending and automating a systems-theoretic hazard analysis for requirements generation and analysis: Massachusetts Institute of Technology; 2013.
19. Rokseth B, Utne IB, Vinnem JE. A systems approach to risk analysis of maritime operations. Proceedings of the Institution of Mechanical Engineers, Part O: Journal of Risk and Reliability. 2017;231(1):53-68.
20. Sulaman SM, Beer A, Felderer M, Höst M. Comparison of the FMEA and STPA safety analysis methods–a case study. Software Quality Journal. 2017:1-39.
21. UK P&I CLUB. Risk Focus: Loss of power. 2015.
22. Leveson N, Thomas J. An STPA Primer. 2015.
23. Leveson N. Engineering a safer world: Systems thinking applied to safety: MIT press; 2011.
24. Blandine A. System theoretic hazard analysis applied to the risk review of complex systems: an example from the medical device industry. Cambridge, MA, USA Massachusetts Institute of Technology; 2013.
25. Becker C, Van Eikema Hommes Q. Transportation systems safety hazard analysis tool (SafetyHAT) user guide (version 1.0). John A. Volpe National Transportation Systems Center; 2014.
26. DNVGL. Guidance for safe return to port projects. DNVGL-CG-00042016.
27. IMO. Formal Safety Assessment - Cruise ships. 2008.
28. Hamann R, Papanikolaou A, Eliopoulou E, Golyshev P. Assessment of safety performance of container ships. Proceedings of the IDFS. 2013:18-26.

8. AUTHORS BIOGRAPHY

Victor Bolbot is a third year PhD student at Naval Architecture, Ocean and Marine Engineering Department of University of Strathclyde, Glasgow. As a PhD student he is conducting research on the safety of complex and Cyber-Physical Systems with focus on power generation systems. His recent research output include publications on safety assessment of power systems on cruise ships, dual-fuel engines and safety assurance methods in Cyber-Physical Systems.

Gerasimos Theotokatos is DNV GL Reader of Safety of Marine Systems at the University of Strathclyde, Department of Naval Architecture Ocean and Marine Engineering. His research focuses on the modelling methods, optimisation and experimental analysis of marine systems and ship energy systems pursuing life-cycle efficiency improvement, their environmental footprint reduction and their safety enhancement.

Evangelos Boulougouris is RCCL Reader of Safety of Marine Operations at the University of Strathclyde, Department of Naval Architecture Ocean and Marine Engineering and Director of the Maritime Safety Research Centre. His main research interests are focused on ships safety and marine design optimisation. He has produced more than 70 publications in journals and international peer-reviewed conferences and 2 chapters in books.

Dracos Vassalos is a Professor of Maritime Safety in the Department of Naval Architecture, Ocean and Marine Engineering at the University of Strathclyde in Glasgow, UK. Professor Vassalos pursued over a 40-year career in industry and academia, promoting the use of scientific approaches in maritime safety and risk, including environmental risk. Professor Vassalos received a Life Achievement Award from the Royal Academy of Engineering in 2011, the Froude Medal from RINA in 2012, the David Taylor Medal from SNAME and a DSC from Strathclyde in 2016 for his life-long contribution to maritime safety.

AGILE POWER MANAGEMENT SYSTEMS – A RULE-BASED CONTROL STRATEGY USING REAL-TIME SIMULATION FOR HYBRID MARINE POWER PLANTS

Dr Chris Watts, Babcock Defence Systems Technology, UK
Dr Truong Quang Dinh, Dr Truong Minh Ngoc Bui and **Dr James Marco,** Warwick Manufacturing Group (WMG), University of Warwick, UK

SUMMARY

With fuel consumption of marine vessels accounting for an increasing portion of global fuel usage, improving the energy efficiency of vessels is essential for saving fuel and reducing greenhouse gas emissions. Hybrid-electric propulsion technologies offer a solution by interfacing multiple power sources, including batteries, with sophisticated energy management systems. An Agile Power Management System for marine vessels is presented by a Babcock led consortium with the University of Warwick (WMG) and Potenza Technology Ltd. The aim of this Innovate-UK funded project is to take advances in automotive energy management techniques and develop a modular marine power management system, addressing the latest guidance and legislation for marine applications. The system employs novel power management algorithms developed using Hardware-in-the-Loop (HIL) modelling techniques. Capable of interfacing energy storage with multiple power sources and loads, the algorithms seek to maximise overall efficiency by improving prime-mover operational envelopes, hence reducing emissions and fuel consumption.

NOMENCLATURE

ABS	American Bureau of Shipping
APMS	Agile Power Management System
BMS	Battery Management System
BSFC	Brake Specific Fuel Consumption
CAN	Controller Area Network
DG	Diesel Generator
DGC	Diesel Generator Control
DP	Dynamic Positioning
EMS	Energy Management System
ESS	Energy Storage System
EV	Electric Vehicle
HEMV	Hybrid Electric Marine Vessel
HIL	Hardware-in-the-Loop
HMI	Human Machine Interface
ICE	Internal Combustion Engine
IMO	International Maritime Organization
KPI	Key Performance Indicator
LTO	Lithium Titanium Oxide
MCA	Maritime and Coastguard Agency
MOL	Maximum Operating Line
MOP	Maximum Output Power
MRBC	Modified Rule Based Controller
NMC	Nickel Manganese Cobalt
OOL	Optimal Operating Line
OOP	Optimal Operating Point
PC	Personal Computer
PDU	Power Distribution Unit
PSV	Platform Supply Vessel
RBC	Rule Based Controller
SOC	State of Charge
WMG	Warwick Manufacturing Group

1. INTRODUCTION

There are ever increasing demands across all industries to reduce environmental emissions and address the global energy crisis by employing low-carbon and energy-saving technologies [1]. The shipping industry is recognised as one of the industries having major impact on global fuel consumption and emissions [2].

As a solution to global energy demands, coupled with the need to reduce emissions, energy management systems (EMS) are high on the agenda across many sectors for further development. The ideal system would increase efficiency, reduce environmental impact and drive down cost; in both production and operation. There is an increasing demand within the marine sector to implement sufficient energy storage into these systems to improve the efficiency by storing surplus energy in times of low load and bolster the supply during a peak load period – improving system resilience.

The development of hybrid technology will be driven in the near future by the IMO climate deal agreed in April 2018, where the shipping industry has committed to reducing emissions by 50% by 2050 compared to 2008 levels [3]. Hence there is a business opportunity to deliver robust and inherently safe solutions to address future demand.

At sea, the vessels deployed on operations are in themselves mobile, condensed versions of the national power grid; this is an area where the whole system can be reinvented by a modern energy system implementing cutting edge energy storage and smart grid technology to improve vessel efficiency and capabilities.

The maritime sector, both military and commercial, is a huge potential market for new energy innovations as the requirements of the modern ship are very different to those of its predecessors. The need for increased functionality, higher levels of automation, longer at-sea life, EU directive compliance and reduced through-life cost can be addressed through the development of an adaptable energy system implementing the latest energy storage and distribution technologies.

Meanwhile, in the automotive sector, recent developments in energy storage technology are increasingly being applied in electric vehicles (EVs). This ranges from mild hybrids, incorporating internal combustion engines with electric motors, to fully electric plug-in battery powered vehicles. There is scope for similar usage in marine vessels, leveraging technology transfer from automotive, to realise several potential benefits:

- Reduced capital cost on electrical generation equipment.
- Improved fuel economy from operating engines in their most efficient power/torque regime.
- Reduced environmental emissions, associated with improved energy efficiency.
- Greater flexibility and robustness from a fully integrated power system.

The introduction of novel energy storage and management systems provides an attractive solution when coupled with modern trends of marine propulsion and power distribution. The development of hybrid and full-electric propulsion, incorporating both direct and indirect drive systems enables greater flexibility, not only in the design of vessels, but also in their operability. This in turn affects through-life cost and environmental impact. The move towards DC distribution facilitates this further, with improved technologies becoming available to overcome the hurdles that have prevented DC adoption over AC distribution in the past [4].

This research seeks to develop a mathematical model to represent a marine power system and a rule-based control algorithm to optimally manage efficient power flow between sources, loads and energy storage. Suitable battery technologies are investigated to fully understand their characteristics and suitability for marine applications. Prototype hardware has been developed, considering requirements of the marine environment, interfacing with energy storage in the form of batteries. This enables the testing and validation of the model and algorithms in a real-time simulation with hardware-in-the-loop.

2. HYBRID VESSEL PROPULSION

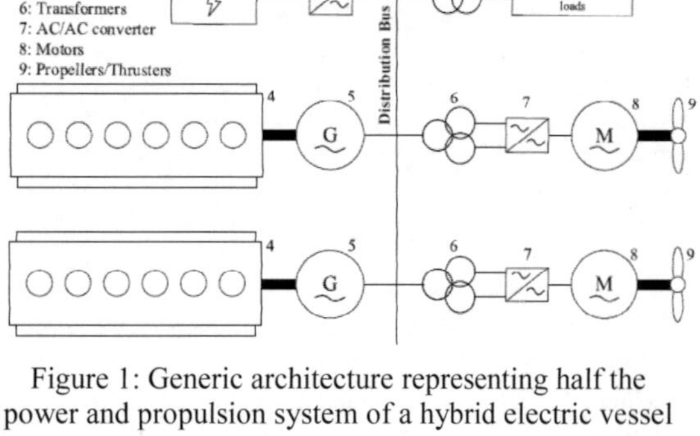

Figure 1: Generic architecture representing half the power and propulsion system of a hybrid electric vessel

The propulsion architecture of a typical hybrid electric vessel is depicted in Figure 1. This effectively represents half the power and propulsion system, which would be replicated for port/starboard, offering necessary levels of redundancy. Generally, the system includes: Power sources, engine-generators and battery, to generate electricity for an AC distribution bus; Transformers and power converters to distribute electricity to propellers, thrusters and auxiliary (hotel load).

This research focuses on platform supply vessels (PSV) as an ideal application of a hybrid power system. PSVs employ dynamic positioning (DP) control systems, which are automated computer controlled systems capable of maintaining the position and heading of a vessel using thrusters and propellers. Due to their advanced control features and accurate positioning capabilities, DP systems are found on a crucial portion of the marine sector, such as cruise ships, cable-laying ships and flotels, as well as PSVs [5].

Operation of a PSV normally follows seven modes: harbour, harbour loading, transit (cruising), DP loading, DP standby, emergency, and black start. For each voyage, the vessel transits to an area desired for operations after loading in harbour at port. It may stay stationary there for a few hours to a few weeks to perform given tasks. The vessel then spends most of its time, and thus energy, in DP loading and DP standby modes (high power demands) before returning to port to unload. In order to ensure the vessel performance in these modes as well as its ability to adapt quickly to any peak loads, the traditional propulsion system needs to have a power capacity much higher than the average power required to operate the vessel. As a result, this leads to the selection of large size engine-generators but running in low power and, therefore, low efficiency regions.

To address the design and control challenges of DP vessels, a hybrid propulsion system using batteries is known as a feasible solution due to its potential benefits. First, the utilisation of batteries allows an optimal control of energy flow between engine-generators and batteries. Depending on load demands and battery energy, one or more engines can be turned off and on to avoid operating in low efficiency zones. This results in significant improvements in fuel consumption, emissions and noise-vibration-hardness level of the engines. Second, similar to automotive applications [6], the use of batteries offers the load levelling function in energy management. By using this function, engine-generators can work at their optimal constant working points with the highest energy efficiency disregarding the actual load fluctuations by dis-charging or re-charging batteries. Third, batteries can provide other advanced features, such as peak shaving (to avoid the need of large power sources to supply peak demands of highly variable pulsed loads, i.e. naval weapons and high-power radar) [7], brake energy recuperation and back-up power [8]. In order to enhance these functions, an agile control scheme is necessary to decide working profiles of engine-generators and batteries in such a way that the fuel consumption and emissions of the engines are minimised while the machine performance is guaranteed [9].

3. DEVELOPING A CONTROL MODEL FOR HYBRID VESSELS

In order to develop the proposed EMS, a backward facing model of the hybrid vessel is constructed based on the following assumptions:

- Transient responses of power electronics and engine-generators are not considered in this study. Energy losses through the power grid are assumed constant and represented by a fixed efficiency value. Meanwhile, energy losses through the engines and generators are represented by an engine break specific fuel consumption (BSFC) map and a generator efficiency map, respectively.
- During simulations, impacts of temperature on the system performance are not considered.
- Total propulsion and hotel loads are represented by a dynamic load profile.

The engine fuel consumption can be derived from a brake specific fuel consumption (BSFC) map versus engine torque and speed as shown in Figure 2. Here, the blue dash line stands for the optimal operating line (OOL) and the red dash line stands for the maximum operating line (MOL) of the engine. In any type of engine, the OOL is the set of optimal operating points (based on engine speed and torque) at which the best engine energy efficiency can be obtained. Likewise, the MOL is the set of operating points at which the engine can provide the maximum output power (MOP).

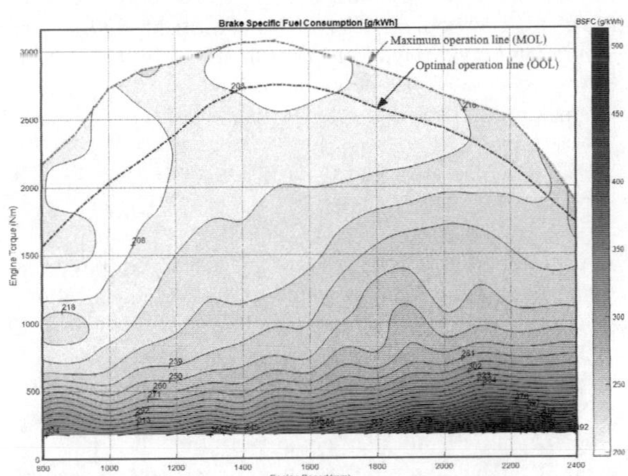

Figure 2: Brake specific fuel consumption map

The instantaneous fuel consumption rate can be calculated as follows:

$$M_{eng} = BSFC\frac{P_{eng}10^{-3}}{0.84} \quad (1)$$

where, M_{eng} is the instantaneous fuel consumption rate *(l/h)*; $BSFC$ is the brake specific fuel consumption *(g/kWh)*; P_{eng} is the output power of the ICE *(kW)*; diesel fuel density is 0.84 *(kg/l)*.

Figure 3 shows the relationship of the power and torque versus speed which represents the performance of the selected engine.

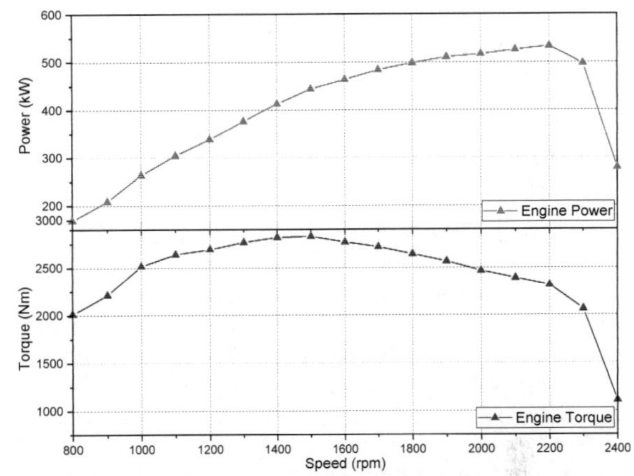

Figure 3: ICE performance map

The control diagram of the proposed EMS for the targeted AC HEMV is depicted in Figure 4. Here, the rule-based EMS block receives the driving command from the load profile block which represents the operational power request. The proposed EMS is constructed based on two rule sets: the logic for operating mode selection and the logic for power split decision. The first set is used to select a proper working mode of the vessel according to load demands. Next, the resulting mode and the current status of the DGs and battery are fed into the second rule set to define the power split between the DGs and battery. The power requests for the DGs are then sent through a procedure to interpolate the torque commands based on their desired speed. The power requests for the battery are converted to the current commands based on the current voltage of the battery. The resulting torque commands and current commands are finally used as the control signals for the ICEs and battery, respectively.

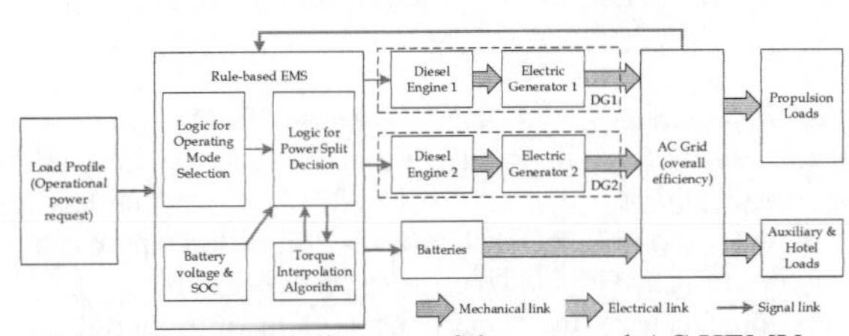

Figure 4: Control diagram of the targeted AC HEMV

The representative load profile based on a PSV is derived as shown in Figure 5. It represents the total propulsion, auxiliary and hotel load demands of the vessel. The load profile comprises of five operation modes: transit (cruising), DP loading, DP standby, harbour loading and harbour.

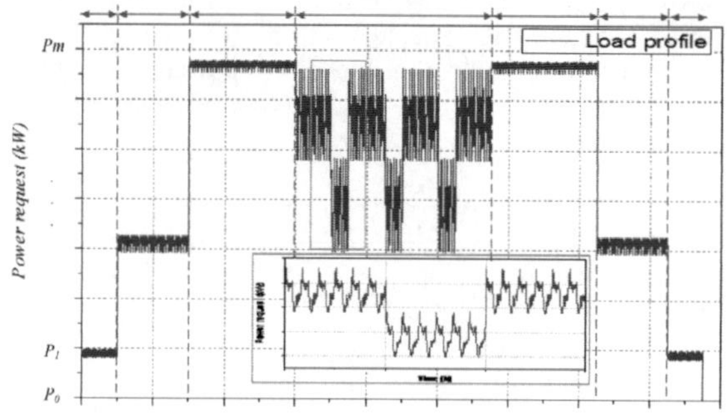

Figure 5: A representative load profile of a PSV

Based on the normal operation characteristics of the selected PSV and the power capacity of the DGs and battery, four power levels are defined for the EMS to allow the system to operate within four different ranges as shown in Table 1 according to the demanded power.

Power request (P_R)	*Power level*	*Vessel mode definition*
$P_R \le P_L$	Very low power request	Harbour
$P_L < P_R \le P_M$	Low power request	Harbour loading + Harbour
$P_M < P_R \le P_H$	Medium power request	DP standby + DP loading
$P_R > P_H$	High power request	Cruising

Table 1: Power level and vessel switching mode definition

In Table 1:

$$\begin{cases} P_L = 0.5\left(P_{ICE}^{opt} + P_{ICE}^{max}\right) \\ P_M = P_{ICE}^{max} + 0.5P_{BAT}^{max} \\ P_H = 2P_{ICE}^{max} \end{cases} \quad (2)$$

where, P_L, P_M, P_H are in turn the low power level, medium power level and high power level of the load demands; P_{ICE}^{opt} and P_{ICE}^{max} are in turn the engine optimal and maximum power at the certain speed; P_{BAT}^{max} is the battery maximum power.

3.1 RULE-BASED CONTROL STRATEGY (RBC)

The RBC strategy is designed for the HEMV based on the power levels of the load demand, and operational mode switching rules as shown in Table 1, respectively. The output power of each DG is controlled up to its maximum power and then the battery depletes the additional energy to support the system or accumulate the excess energy from the system. The battery pack representing the ESS plays an important role in optimizing the usage of the DGs. The following rule-based logic is employed in the RBC in order to achieve the control target.

Rule 1: If P_R is less than or equal to P_L

Then, if the remaining energy in the battery is sufficient to support the system within a pre-set duration, the two DGs are kept OFF and the battery is depleted to support the system; else, when the battery energy is lower than 75% SOC, one DG is started and kept to operate at the MOP. In this circumstance, the battery accumulates the excess energy from the system for later usage.

Rule 2: If P_R is larger than P_L and less than or equal to P_M

Then, only one DG is started and kept running so that the output power is up to MOP. The battery is switched to charging or discharging mode according to its required power and SOC status. In this case, the battery SOC is limited so that it can be charged up to 95% or discharged up to 70% to maintain the performance of the battery.

Rule 3: If P_R is larger than P_M and less than or equal to P_H

Then, depending on the load demand and current status of battery SOC, one DG or two DGs are turned on and kept running so that the DG's output power is up to MOP. The battery SOC's thresholds are set the same as in Rule 2. When the battery SOC reaches the bottom threshold, it is switched to charge mode and both DGs start. In this circumstance, the output powers of both DGs are kept the same to balance the performance of the two DGs.

Rule 4: If P_R is larger than P_H

Then, the two DGs are turned on and hold running at MOP and the battery is discharged to a defined target threshold. When the battery SOC reaches the bottom threshold, the battery is switched off and there are now only two DGs operating at MOP. The maximum output power of the system in this case can be estimated by the total power generated by two DGs and the maximum discharging power of the battery.

3.2 MODIFIED RULE-BASED CONTROL STRATEGY (MRBC)

As developed from the RBC, the MRBC employs the same rules of the RBC as stated in the previous section. However by using the load levelling concept with the improvements to overcome the issue of the RBC, the ICEs are managed to operate at or close to the OOP following the OOL to achieve the highest energy efficiency. The torque command for the ICE is interpolated based on ICE's power request and its desired speed so that the output power of the ICE is close to OOP before requesting power from the battery.

4. BATTERY CHARACTERISATION TESTING

In the development of a prototypic hardware and energy storage solution, numerous battery chemistries and physical formats were considered. In order to set-up the battery management system (BMS) optimally, the characteristics of the cells being used are required.

Characterisation of cells in terms of fully understanding their performance under different operating conditions is important when considering the service life of the cells. It is also important when the cells come to the end of their first-life application and can then be considered for second-life, where re-characterisation is used to sort cells for suitable second-life applications. This is particularly important for a sustainable approach to energy storage systems across all sectors, not just marine. It is also of great interest to through-life support provision for such systems and recognising the opportunities for second life applications as part of the service delivery to the end-user.

There are many characterisation standards that are applicable to battery systems, primarily driven by automotive guidance and legislation. These standards apply to different geographical areas of the world at either cell or pack level and can overlap and contradict in some instances. In the marine sector, legislation and guidance is still relatively immature, with battery packs/systems being approved on a case-by-case basis as the technology is rapidly developing.

The battery characterisation testing undertaken during this work was performed with the consideration of these wide-ranging standards.

The testing sought to quantify key performance measures representative of a marine environment, such as:

- Cell capacity at various temperatures and current rates
- Cell impedance at various temperatures and current rates
- Cell durability under mechanical and thermal loading

Consideration was given to different use cases appropriate for the marine environment:

- Operating requirements (e.g. usage cycles)
- Environmental conditions (e.g. temperature, humidity)
- Robustness and reliability constraints

Cell chemistries and formats that were established as being of interest for future marine applications were identified based on the requirements of the aforementioned standards, safety considerations, pack integration, service lifetime and performance characteristics. The following cells were chosen for characterisation testing:

- Lithium Nickel Manganese Cobalt Oxide (NMC), 18650 format – Sony VTC6
- Lithium Titanate (LTO), pouch format – Altairnano
- Supercapacitor – Maxwell

The test programme was carried out in the purpose built laboratory at WMG. An example of some of the results obtained can be seen in Figure 6.

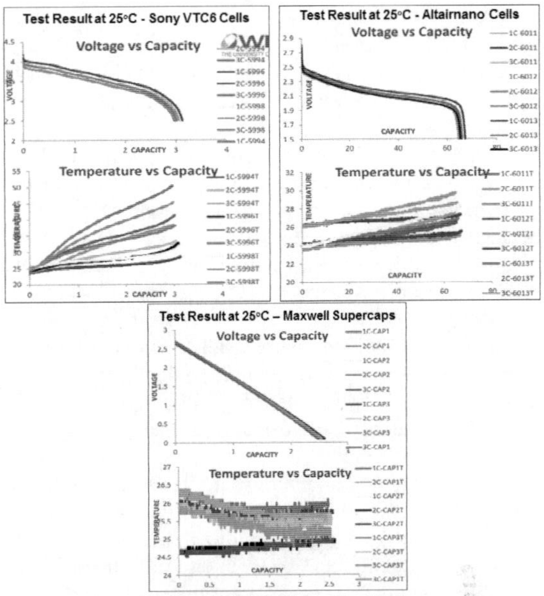

Figure 6: Cell Capacity Tests

Based on integration, handling, overall performance and characterisation data, the Sony VTC6 cells were chosen for the prototype battery module. The 18650 format of these cells is familiar to the automotive industry, and thus Potenza Technology Ltd, and makes them easy to handle and integrate into a battery pack. The characterisation data was fed into Potenza's battery management firmware to enable optimal interfacing with the control algorithms and marine energy management model developed by WMG.

5. PROTOTYPE HARDWARE

A prototype power management system incorporating a battery module has been designed and built by Potenza Technology Ltd. The requirements of this hardware were established by adopting automotive standards and addressing current rules, legislation and guidance provided by Lloyd's Register, DNV GL, MCA and ABS for installation of batteries and battery management systems on-board marine vessels.

The requirements of the system consider the following key criteria:

- Hazards (mechanical, electrical, chemical, fire, environmental)
- Location on a marine vessel (space envelope and configuration)
- Environmental control (temperature, humidity, cooling and ventilation, hazardous area designation)
- Fire safety (detection, prevention, suppression)
- Management system (battery management, monitoring, interfaces, communications, software)
- Testing (at cell and system level, environmental, interfaces, alarms, safety functions)
- Operation and maintenance (human factors, through-life management and supportability)

The prototype incorporates: a battery management system (BMS); power distribution unit (PDU); communications interfaces (using CAN); a battery module housing the Sony VTC6 cells; and a HMI for test operations.

The components of the unit are housed in a standard 19" rack mounting system, with all interconnects internal to the cabinet, allowing a "plug-and-play" approach for the modules. The battery module is rated at 43V with a capacity of 90Ah and peak power input and output of 7.5kW and 15kW respectively. It also features local cell monitoring and balancing, fusing for overcurrent/short circuit protection and a pressure relief safety system. The battery module is designed for single person handling.

The prototype BMS and battery unit can be seen in Figures 7 & 8 alongside the real-time hardware-in-the-loop (HIL) test equipment.

The prototype aims to demonstrate interfaces with a marine vessel power system including loads (i.e. propulsion, hotel), sources (i.e. diesel generators) and enables full testing and validation of the energy management model and control algorithms.

6 EXPERIMENTAL TESTING

6.1 REAL-TIME SIMULATION USING HIL

Real-time simulation and HIL platforms have been widely used for many years across numerous sectors, primarily automotive, for rapid prototyping and system development. These methods have been adopted by the marine sector, where they facilitate rapid validation of control systems, including complex power management, in a safe environment.

A real-time simulation platform allows a physical system to be represented by an equivalent model running on hardware that interfaces with prototype physical equipment and control systems. This enables real-world working conditions to be replicated and allows validation of the system model and control logic in real-time. Hence, this approach offers a cost-effective, fast and safe way of testing new systems and control methodologies prior to deployment in the field. In this context, dynamic closed-loop testing of energy storage systems can be performed that is deterministic and repeatable.

For this study, the aim was to set up a real-time HIL platform in order to fully test and validate the marine propulsion system model and algorithms that have been developed. A bespoke optimisation framework for marine energy management has been designed and implemented, considering interfaces with the physical prototype BMS unit and battery module.

6.2 REAL-TIME HIL CONFIGURATION SETUP

A real-time simulation platform has been built in the Control Laboratory at WMG, the University of Warwick, using equipment from dSPACE. In this study, the HIL platform is set up as shown in Figure 7. The system consists of a SCALEXIO which is used to simulate the propulsion model and the EMS; the prototype battery module (43V, 90Ah) including the BMS to manage the battery and represent the energy storage system; and a Bitrode module cycler which is used to charge/discharge the battery, thus representing power sources and loads. To facilitate communication between the hardware, a common controller area network (CAN) bus is employed. A Vector CANalyzer device is installed to monitor all the CAN messages for monitoring purposes while all system operating data is logged using dSPACE ControlDesk software running on a PC. Figure 8 displays the actual test bench of the HEMV.

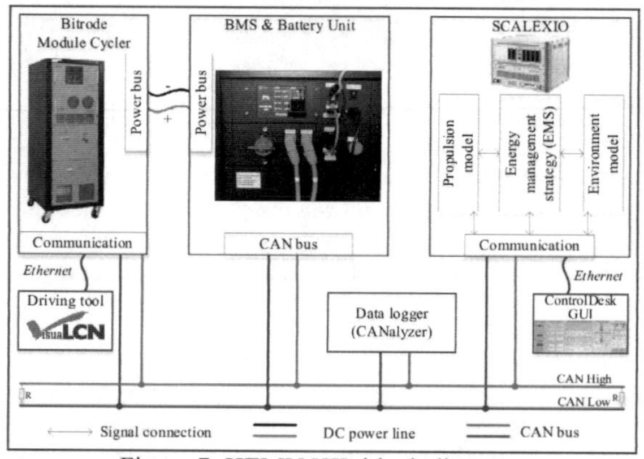

Figure 7: HEMV HIL block diagram

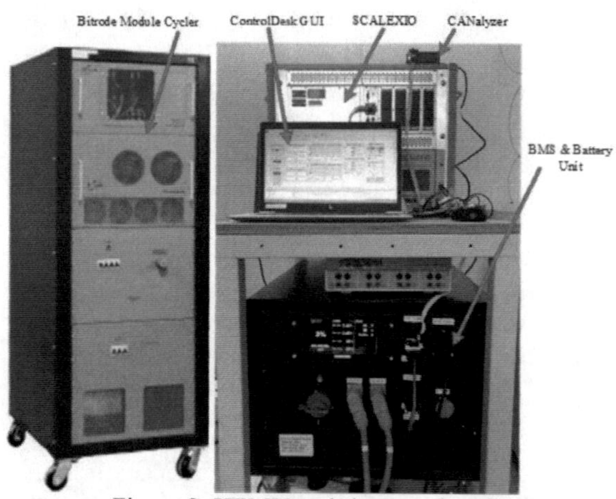

Figure 8: HEMV real-time test bench

7. REAL-TIME SIMULATION RESULTS

The testing of the whole HEMV was carried out using the above HIL setup to evaluate the control performance of the proposed MRBC on the real battery module. The results can be used to assess the improvement of energy cost and fuel consumption and thus the emissions of the HEMV by real-time testing. The load profile with five operation modes shown in the previous section is used as the power demand of the voyage. The total duration of the load profile is scaled to 17.5 hours to enable testing in laboratory conditions. The DGs start/stop duration is set to 20 minutes and the testing starts when the initial SOC is 85%.

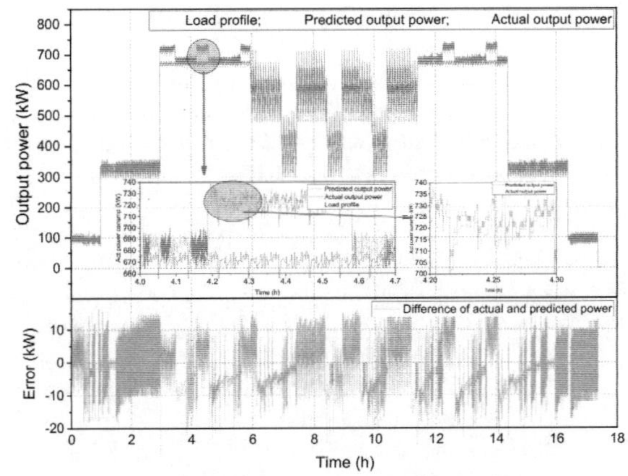

Figure 9: HEMV output power

Figure 9 demonstrates performance of the HEMV in real-time under the MRBC control strategy. The top sub-plot shows the grid power request by the load profile and the predicted output power that the diesel-generators and the battery have to generate. It can be seen that the predicted output power is always higher than the power request. This is because of the performance efficiency of the DGs, Coulomb efficiency during charge/discharge cycle of the battery and the whole powertrain and grid efficiency. The top sub-plot also presents the power comparison between the predicted and actual output power of the HEMV in real-time testing. It can be seen that those output powers fit very well with each other, in that they have the same operating form. The bottom sub-plot shows the power difference between the predicted and actual output power.

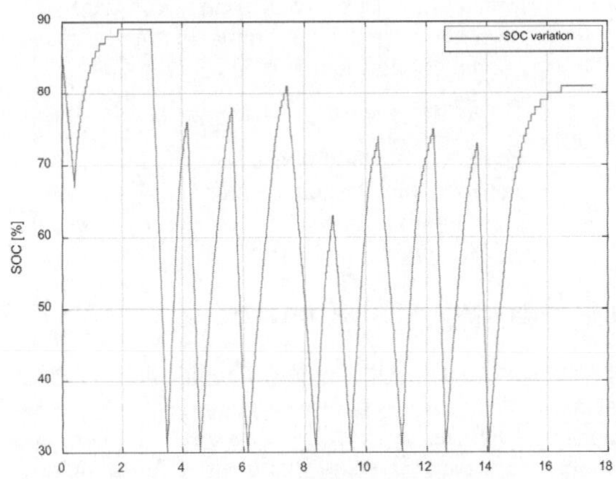

Figure 10: Battery state of charge variation

Figure 10 shows the variation of the battery SOC over time during the journey. As mentioned in the previous section, the MRBC determines power split and power commands for the DGs and the battery based on the load levels. Therefore, the battery in this circumstance is controlled to be charged or discharged during the voyage depending on the status of the SOC and the actual power command. Here, the battery SOC variation satisfies the safety constraint as it varies within the range between 30% and 90% of SOC.

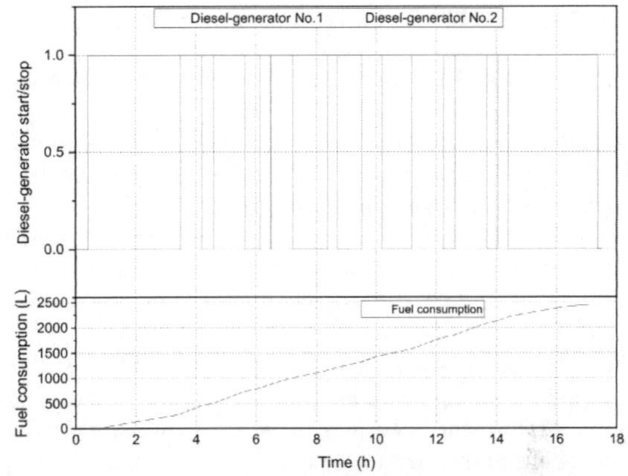

Figure 11: Diesel-generator start/stop sequence and total fuel consumption

The top sub-plot of Figure 11 shows the start/stop sequence of the DGs while the bottom sub-plot shows the total fuel consumption over time during the journey. The minimum start/stop duration satisfies the constraints of this test-case (that is the ICE should run for at least 20 minutes before stop).

To demonstrate the performance improvement of the proposed MRBC over the conventional DG control strategy (DGC), the key performance indicator (KPI) table is built based on the control design requirements as illustrated in Table 2. The KPI table enables the comparison of the performance of the HEMV system between control strategies in terms of ICE start/stop cycle, operation time per voyage, fuel consumption and energy cost. It is noted that the vessel energy cost is the total cost of fuel consumption, electric cost for cranking the ICEs and electric cost which produces the difference between the initial and final battery SOC.

KPI	Conv. Non-hybrid Vessel (DGC)	Hybrid Vessel (MRBC)
Total Run Time [hh:mm:ss]	17:30:00	17:30:00
Gen 1 Stop/Start Count [-]	1	1
Gen 1 Run Time [s]	63000	61051
Gen 1 Stop Ratio [%]	0	3.09
Gen 2 Stop/Start Count [-]	13	10
Gen 2 Run Time [s]	37581	22670
Gen 2 Stop Ratio [%]	40.35	64.02
Battery SOC Drop [%]	-	4
Vessel Fuel Used [L]	2576.2	2457
Vessel Fuel Cost [£]	3233.1	3083
Vessel Elec Used [kWh]	-	9.61
Vessel Elec Cost [£]	6.94	6.64
Vessel Energy Cost [£]	3240	3089.7
Energy Cost Saved vs DGC [%]	-	4.6%

Table 2: KPI comparison of the two control strategies

From Table 2, it can be seen that the total fuel consumption of the hybrid vessel is reduced by 119.2*l* (equivalent to 4.63%) using the MRBC when compared to the conventional non-hybrid vessel using the DGC. Overall, the MRBC achieves a 4.6% saving of the total energy cost. In addition, under the MRBC, the DGs can be turned off to save energy. The stop ratio of DG1 is 3.09% while that of DG2 is 64.02%. When the DGs are turned off, the battery is the main power source to support the system.

8. CONCLUSIONS AND FURTHER WORK

This study investigated an intelligent power management system interfacing with a simulated hybrid electric propulsion system for marine vessels.

Battery characterisation data was obtained and fed into a prototype BMS to optimally interface with a physical battery module and the simulated hybrid marine vessel model.

The control strategy based on the RBC was successfully employed to drive the system in such a way to maximize engine performance and to achieve the benefits in terms of fuel consumption and emissions.

The effective EMS was introduced for a case study on HEMVs. Real-time HIL simulations were performed to verify the basic functionalities and performance of the designed EMS compared to a conventional non-hybrid vessel. This demonstrated that significant fuel and overall energy cost savings can be realised over a relatively short period of time.

Future work will seek to further develop the energy management algorithm, such as to increase the adaptability of the MRBC through optimising the weighting factors and/or to consider other factors (e.g. system cost and emissions), in order to improve system performance. Further validation of the proposed strategy in the real-time environment will be considered by employing several different real-world load profiles and usage cycles to verify its practical applicability.

Furthermore, the limit of engine stop-and-start rate (i.e. per voyage/day) needs to be taken into account when performing the optimisation to minimise its negative impacts on long-term performance and accommodate necessary levels of spinning reserve.

The opportunities for through-life support of battery systems for marine applications will be investigated further, considering the exploitation of battery second-life applications in order to benefit end-users and the circular economy.

9. ACKNOWLEDGEMENTS

This research was supported by Innovate UK through the Agile Power Management Systems (APMS) project, in collaboration with Babcock International Group, WMG at the University of Warwick, and Potenza Technology Ltd.

10. REFERENCES

1. HM Government, 'The Carbon Plan: Delivering our low carbon future', *DECC, London*, 2011.

2. International Maritime Organisation (IMO), 'Third IMO greenhouse gas study 2014: executive summary and final report', *Technical report, London (UK)*, 2015.

3. IMO, 'Briefing: 06 - UN body adopts climate change strategy for shipping', http://www.imo.org/en/mediacentre/pressbriefings/pages/06ghginitialstrategy.aspx, April 2018.

4. BSI. 'BS8877-1:Design for manufacture, assembly, disassembly and end-of-life processing (MADE) - Part 1', *Br Stand Inst*, 2006.

5. DP & Marine Assurance Norway AS, 'Dynamic Positioning', http://www.dpmarine.org/services/dynamic-positioning/ , August 2018.

6. Son H., Kim H., Hwang S. & Kim H., 'Development of an Advanced Rule-Based Control Strategy for a PHEV Using Machine Learning', *Energies, vol. 11(1), pp. 1-15*, 2018.

7. Dedes E.K., Hudson D. & Turnock S.R., 'Assessing the potential of hybrid energy technology to reduce exhaust emissions from global shipping', *Energy Policy, vol. 40, pp. 204-218*, 2012.

8. Geertsma R.D., Negenborn R.R., Visser K. & Hopman J.J., 'Design and control of hybrid power and propulsion systems for smart ships: A review of developments', *Applied Energy, vol. 194, pp. 30-54*, 2017.

9. Sun J., 'Optimization-based control for electrified vehicles: challenges and opportunities', *Journal of Control and Decision, vol. 2(1), pp. 46 – 63*, 2015.

11. AUTHORS' BIOGRAPHIES

Dr Chris Watts is Technology Acquisition Lead at Babcock Defence Systems Technology. He is a Chartered Engineer and member of the IET. He obtained his PhD from the University of Leicester in the field of brushless motors and drives. He is currently working within R&D to expand the company product portfolio into new markets.

Dr Truong Quang Dinh is Assistant Professor in Energy Management and System Control at WMG, University of Warwick. His research interests include;

advanced mechatronic systems, system engineering, power management, control theories and applications.

Dr Truong Minh Ngoc Bui is a Research Fellow at WMG, University of Warwick. His research interests include real-time control, power management and battery management in hybrid electric systems.

Dr James Marco is a Reader in Vehicle Electrification and Energy Storage at WMG, University of Warwick. He is a Chartered Engineer and member of the IET. His research interests include the application of systems engineering, real-time control, and physical systems modelling for low carbon technologies in the transport and energy storage sectors.

INVESTIGATION OF AUXILIARY POWER POTENTIALS OF SOLAR PHOTOVOLTAIC APPLICATIONS ON DRY BULK CARRIER SHIPS

Wandifa Saidyleigh, Gambia Maritime Administration

A I Ölçer, R Baumler, World Maritime University, Sweden

ABSTRACT

The increase in world seaborne trade over the past decade due to global economic and population expansion has resulted in a corresponding increase of world shipping fleet with even greater size and power requirements. The bulk of these ships use cheap and widely available fossil fuels, mainly oil for operation but which has deleterious effects on the environment. In order to address environmental concerns in the shipping sector, the International Maritime Organization (IMO), responding to the global call to reduce greenhouse gases emissions from international shipping adopted technical and operational measures. These are to ensure efficient energy management on ships and have led to the application of many innovative technologies including the use of renewable energies and alternative fuels on ships to minimize fossil fuel consumption and reduce emissions.
However, in order to achieve a substantial emissions reduction in international shipping, the potential applicability of a technology which utilizes a universal renewable energy resource on the largest ship type in international shipping fleet should be investigated. This research focuses on investigating the potential of Solar Photovoltaic technology on dry bulk carriers using a developed methodology and Levelised cost of energy concept as the basis for comparison. The results of this research can be used to guide decision makers about the potentials of Solar Photovoltaic technology on dry bulk carriers in general whilst its developed methodology may be useful in the specific context for determining which ships and under what circumstances solar PV is an option.

KEY WORDS: Bulk Carrier, CO_2 Emissions, Engine load, Fuel Consumption, Levelised Cost of Energy, Photovoltaic, Simulation

1. INTRODUCTION

Maritime transport over the last decade has seen huge increase in the global fleet volume mainly as a result of the growth in the world trade [16]. The impact of the fleet growth has been negatively felt from the increasing level of GHG emissions due to the continued dependence on fossil fuels [9]. IMO's reaction to curb the scenario increase in the emissions trend recently came through regulatory framework encompassing the technical and operational measures. These measures have become drivers for innovation within the international shipping industry in the search for more energy efficient systems including abatement technologies. Whilst much effort and time has been dedicated to investigating the potentials of certain renewable energy technologies such as wind, hydrogen fuel and biofuels, in comparison, the search for solar photovoltaic (PV) potentials in scope and application in ship energy systems has been quite limited. Although studies have been carried out into the potentials of solar energy applications on certain type of vessels and its actual applications on many different vessels in service, huge potentials remain unexplored. [11] researched into the optimal sizing of PV/diesel/energy storage system on a 100,000dwt oil tanker. In a similar research [13] carried out on a research boat to explore the PV technology hybridized with diesel engines have both revealed huge potentials in terms of fuel savings, and the possible cost and emissions reductions. In actual applications on pleasure crafts 'Solar Sailor' [14] and MS Turanor Planetsolar, where PV serves as the sole power source to the provision of auxiliary power on a car carrier where it accounts for about 7% of the power [5]. In spite of these developments, gaps still exist with respect to PV application potentials on specific type of vessels such as dry bulk carriers. Dry bulk ships constitute 43.1% of international shipping deadweight [17] and the second highest CO_2 emitter next to container vessels [15]. Therefore, a huge impact could result in terms of emissions reductions, and fuel cost savings when the potential of PV technology is explored for this ship type.

The motivation for this research is premise on the fact that vast spaces or areas exist on a dry bulk carrier hatch covers on the exposed deck which remain economically unutilized from cradle to grave in the life of the ship. Solar power deployment is quite dependent on the availability of space or area which is in direct contact with the solar radiation as well as the efficiency of the module. [14] indicated that if $9m^2$ of photovoltaic panels are required to produce 1kw of electrical power, an area of 2,700 to $8{,}100m^2$ corresponding to the most frequent auxiliary power range of 300-900kw for merchant vessels will be required. Secondly, photovoltaic installation can coexist with other innovative technologies such as Fletner rotor and kites on deck to provide additional auxiliary power for the ship and does not pose safety threats to crew access routes.

In order to establish the potentials of photovoltaic technology on dry bulk carrier, this research will utilise

Levelised Cost of Energy (LCOE) concept as basis for comparison between a Conventional (existing) case and a Hybrid case. The Conventional case utilises only diesel engines to provide electrical energy for auxiliary purposes whilst the Hybrid case utilises diesel/solar combination to provide same electrical energy under varying electrical load of ship. However, since a ship's auxiliary diesel is designed to endure maximum load as well as surges due to transient loads, comparison on the basis of LCOE in this research takes in to account only the average electrical load demand.

LCOE is the usual metric or measurement standard to compare the costs of different electricity generation technologies [8]. Levilised energy is a ratio of annualised total cost to the total energy produced by the system during one year period. The total cost which is the sum of capital expenditure (Capex) and operational expenditure (Opex) is annualised by factoring the capital recovery or annualised factor in the total cost. [3] defined it as the ratio of the sum of discounted lifetime generation costs to the sum of discounted lifetime electricity output. The specific cost elements considered within the scope of this research will be dealt with in the subsequent sections.

In determining the LCOE, this research considers the cost elements associated with the production of electrical energy in both conventional and hybrid cases using average auxiliary engines load situation. The rationale for choosing average load on auxiliary engines is predicated on possible electrical load variations in different ship operating modes such as at sea, manoeuvring, berth and anchorage [6].

2. PROPOSED METHODOLOGY

This research was constrained by the lack of real time ship operating data due to its proprietary and sensitive nature. Alternatively, the author is proposing a methodology to determine average values of the parameters associated with operational cost elements based on the available engine technical data and then apply Monte Carlo simulation using Oracle Crystal Ball to obtain probabilistic values. These parameters are average daily fuel consumption and average engine load. The author will rely on secondary data obtained from Clarkson's Database for use in the case study.

As ships are invariably fitted with diesel engines to provide electrical energy for auxiliary purposes, this is considered as the conventional case, whereas the case of the diesel/solar is considered in this research as the hybrid case.

During the author's review of the available literatures on the specific applications of Solar PV on ships, it was discovered that the significance of those studies were limited to the potential fuel savings and associated economic benefits to the particular ship. Those studies did not utilize LCOE to compare energy production systems on a number of ships of same type and they also fall short of developing a methodology to determine the average fuel consumption which may serve as a decision support tool for new builds. This is the uniqueness of the author's approach.

This research will investigate the auxiliary power potential of Solar PV application on many bulk carriers of different sizes by utilizing the empty spaces on the hatch covers and draw comparison with conventional cases of same ship sizes based on LCOE concept. The research will also attempt to demonstrate the potential positive impact of the hybrid case on the environment through reduction of greenhouse gases.

Schematic Presentation of the approach is given in figure 1 below.

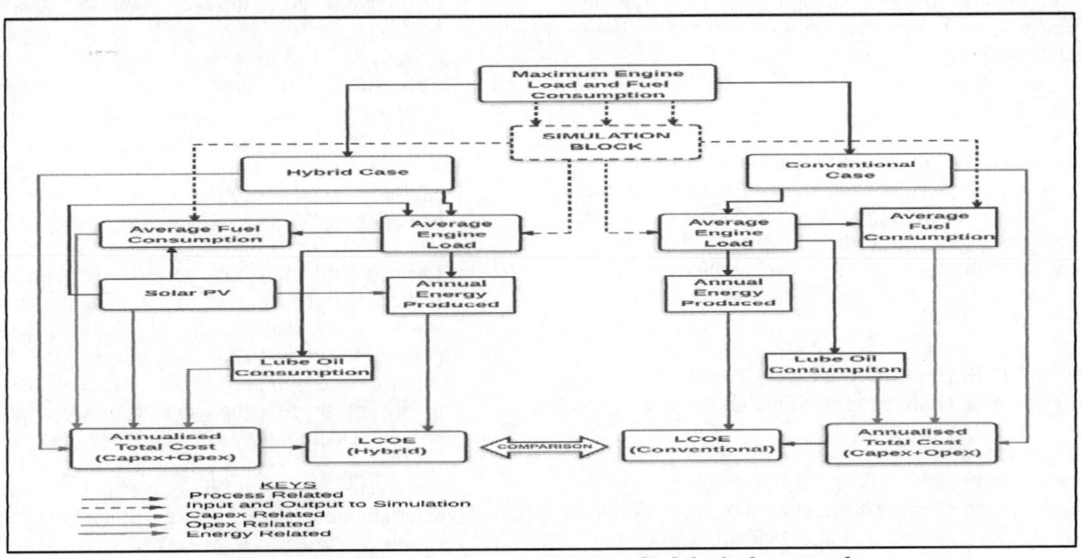

Figure 1: Schematic diagram of the methodological approach

2.1. MATHEMATICAL MODELLING OF THE ESSENTIAL PARAMETERS

2.1.1 AUXILIARY ENGINE LOAD AND SHIP ELECTRICAL LOAD

Auxiliary diesel engine(s) in a conventional ship provides the entire electrical load since it's the only power source available to fulfil ship's electrical power requirements. Electrical power demand of a ship at any time determines its auxiliary engine load capacity and that depends on ship's operating mode [15]. Therefore

$$Electrical\ Load\ (\text{max}) = f * Engine\ rating$$

$$\boldsymbol{E_{max} = P * f} \qquad \text{Equation 1}$$

Where E_{max} *is the ship's maximum electrical load (kw) under service condition, P is the total engine power rating (kw) of the auxiliary engines and* f_{max} *is the load factor.*

For operational efficiency, diesel engines are constrained to operate below certain minimum threshold load of maximum continuous rating [12] determined by the engine builder, and ship's minimum electrical load therefore corresponds to this value.

Therefore, minimum electrical load on the ship due to minimum load factor of f_{min}.is given by

$$\boldsymbol{E_{min} = P * f_{min}} \qquad \text{Equation 2}$$

2.1.2. MODELLING OF FUEL CONSUMPTION BASED ON THE AUXILIARY ENGINE LOAD

Fuel consumption is a function of the engine power, and in the conventional case considered in this research, all electrical loads is supplied from the combustion of fuel in the diesel engine. This can generally be determined from equation 3 below.

$$\boldsymbol{Q_t = S_f * P * f * t} \qquad \text{Equation 3}$$

Where Q_t (tons) is the amount of fuel consumed in time t (hrs) and S_f (g/kwh) is the specific fuel oil consumption of the auxiliary diesel engine. S_f (g/kwh) for high speed diesel engines varies according to engine technology, characteristics and age but range generally between 185-225g/KWh for the technology period pre-1993 to post 2001[6].

2.2. DETERMINATION OF AVERAGE POWER AND AVERAGE DAILY FUEL CONSUMPTION

Ship's auxiliary diesel engines are designed to operate at different loads to satisfy changing power demand of the ship board electrical equipment. How frequent and long auxiliary engines are operated at or below rated power is complex to determine as this depends on changing electrical power demand. This situation practically noticed during manoeuvring when start air compressors, and steering gear pumps use to run quite often whilst hydraulic pumps for winches, windlasses and cargo gear lifting equipment are used more often at berth. It is therefore reasonable to determine average power and average daily fuel consumption. In the Third IMO GHG Study, auxiliary loads were estimated based on the number of engines and rated engine power, in this study, it is determined from equation 4 below.

$$\mathbf{P}_{av} = \frac{1}{n}\sum_{t=0}^{t=k} \mathbf{P}_t \mathbf{f}_t \qquad \text{Equation 4}$$

Where P_{av} is the average power of an auxiliary engine in operation corresponding to the nominal electrical load demand of the ship, P_t is the total power of the auxiliary engines at time (t), f_t is the load factor at time (t) and n is the number of auxiliary engines in operation. The interval t=0 to t=k is the period during which this load condition exists.

Similarly, the average consumption can be determined as follows;

$$\mathbf{Q}_{\mathbf{fav}} = \frac{1}{n}\sum_{t=0}^{t=k} \mathbf{Q}_t \qquad \text{Equation 5}$$

Where Q_t is the fuel consumption in time (t), n is the number of auxiliary engines in operation, Q_{fav} is the average daily consumption of an auxiliary engine.

The determination of these average values is necessary for the following reasons:

I. The electrical load on the auxiliary engine is proportional to the engine load and its varies continually depending on the service condition of the ship as well as shipboard activities

II. Total auxiliary power available on the ship and load factor are variable quantities; this is because a ship may have many auxiliary engines such as 2, 3 and in some cases up to 4 but only some would be required to operate depending on the load situation such as ballast condition and other auxiliaries engine may be started as ship service loads increase. Even those running auxiliary engines may not be loaded to their full power rating for the same reason.

III. Existing ships may have records of operational fuel consumption data of the existing auxiliaries to use as reference but this is not the case for new ships. In that case it would be useful to determine consumption from average values of power and specific fuel consumption instead of absolute value of the engine power rating.

IV. The average power values are important in the determination of total energy produced by the auxiliary engine in a year as will be shown in the data presentation chapter. Using the rated auxiliary power for annualised energy will give a bloated result.

Since values of average load and average daily fuel consumption are probabilistic in nature determined from varying situations explained earlier, the author utilizes Monte Carlo Simulation as the analytical tool to produce forecast values of P_{av} and Q_{fav} from assumption and decision variables. These will be explained in more detail in the subsequent section.

Monte Carlo simulation using Crystal Ball produces probabilistic values of a model whose output depends on the effects of the varying inputs.

2.3. SOLAR POWER DETERMINATION

Solar power is determined only in the hybrid case where part of the electrical load would be provided by Solar Photovoltaic cells installed on the hatch covers in contrast to the conventional case. Key factors to consider in solar power production are the area, *A* available to solar irradiance, intensity of the global solar irradiance, *G* and efficiency, η of solar module. Whilst performance of PV modules are determined under specific set conditions called Standard Testing Conditions (STC) [20] and that PV performance is influenced by environmental factors [4], the effects the later has not be considered in this research. The rationale for neglecting environmental factors is due to the fact that field measurements are required in order to accurately determine the performance of PV system [4].

Power generated from PV system can be determined generally from equation 6 below

$$P_s = \eta_s * A * G \qquad \text{Equation 6}$$

Where P_S is the power (w) produced by the system at the averaged global irradiance of G w/m^2 per day, η_s the efficiency of the solar module and *A* is the installation area in m^2. The area considered in this research is total hatch area of the bulk carrier equivalent to the installable module area of the PV system.

2.4. ELECTRICAL LOAD OF THE AUXILIARY DIESEL ENGINE IN THE HYBRID CASE

As auxiliary diesel engine is the main source of power and the solar PV is only complimenting, in the hybrid case, the electrical load can be determined from equation 7 below:

$$ELd_h = P_{av} - P_S \qquad \text{Equation 7}$$

Where $ELd_h$is the electrical load (balance) of the auxiliary diesel engine when the Solar PV system supplied electrical load P_S. It should be noted that P_S is available for only 12 hours, which is the average length of daytime the period when solar energy is expected to be available.

2.5. DETERMINATION OF FUEL CONSUMED IN THE HYBRID CASE

As the power (load) of the diesel engine is the hybrid case is lower than that in the conventional case by an amount equivalent to P_S, the consumption in the hybrid case therefore can be determined from equation 8 below:

$$Q_h = Q_{fav} - 12 * P_S * S_f \qquad \text{Equation 8}$$

Where Q_h is the average daily fuel consumption in the hybrid case, S_f is the specific fuel consumption and P_S is the available generated power from Solar PV for 12 hours period.

2.6. DETERMINATION OF LUBRICATING OIL CONSUMPTION IN CONVENTIONAL AND HYBRID

The amount lube oil consumed depends on the engine load, if S_L is the specific lube oil consumption in g/kwh, for the two cases consumption (g/hr) is given by:

$$C_L = S_L * P_{av} \qquad \text{Equation 9}$$

$$H_L = S_L * Eld_h \qquad \text{Equation 10}$$

Where C_L and H_L are lubricating oil consumptions for Conventional and Hybrid cases at the respective average engine loads.

2.7. ECONOMIC ANALYSIS

In order to determine LCOE, Total Cost (*TC*) and Total Energy Produced (*TEP*) in both conventional and hybrid cases has to be established.

2.7.1. COST ELEMENTS IN THE CONVENTIONAL CASE

i. Capital cost of the auxiliary diesel engine(s) referred as Capex

ii. Cost of fuel consumed for one year period

iii. Cost of Lubricating Oil consumed for one year period

2.7.2. COST ELEMENTS IN THE HYBRID CASE

i. Capital cost of the auxiliary diesel engine(s) Capex

ii. Cost of fuel consumed for one year period

iii. Cost of lubricating oil consumed for one year period

iv. Capital cost of Solar PV system-Capex this excludes the Balance of System components (BOS) such as battery, inverter and associated wiring.

In the conventional case ii and iii are the components of Opex and in the hybrid case, ii, iii, iv are the components of Opex. Owing to lack of reliable data on costs associated with auxiliary engine maintenance such as spares and engine cooling treatment which differs according to builder's recommendations, these cost elements have disregarded in this research. In a similar manner, cost associated with PV system maintenance is also neglected.

2.8. DETERMINATION OF LEVELISED COST OF ENERGY

The Levelised cost is given by the relationship below equation 11 below:

$$\mathbf{LCOE} = \frac{TC*f}{TEP} \qquad \text{Equation 1}$$

Where LCOE is the Levelised cost of energy in $/KWh, *TC* is the sum of Capex and Opex, *TEP* is the total energy produced by the system over a year period, and *f* is the Capital Recovery or annualised Factor, CRF. This factor converts the initial cost of investment to annual cost and is given by equation 12 below:

$$\mathbf{CRF} = \frac{r(r+1)^n}{(r+1)^n-1} \qquad \text{Equation 12}$$

Where r is the interest rate and n is the life span of the system.

3. CALCULATION AND DATA PRESENTATION

For the purpose of case study, secondary data was obtained from Clarkson's database consisting selected categories of 10 dry bulk carrier ships whose hatch dimensions were available in the database. Ships selected came from three main size categories according to deadweight. These are Capsize, Panamax and Handymax constituting 70% of the total dry bulk carriers in the database at the time of selection. Handysize ships with dimensions were not available.

3.1. SELECTION OF SHIPS IN THE CATEGORIES

Capesize 100,000 Dwt and above: This represents 15% of the total bulk carriers in the database. Four Ship 1, Ship 2,& Ship 3 in decreasing order of deadweight are selected from this category.

Panamax-65,000-99,999 Dwt: This represents 23 % of the total bulk carriers. Four ships are selected from this category and represented in decreasing order of size, Ship 4, Ship 5, Ship 6 and Ship 7

Handymax-40,000-64,999 Dwt: This represents 32% of the total bulk carriers in the database. Three ships are also selected and numbered in decreasing order of size, Ship 8, Ship 9 and Ship 10. The raw data obtained from Clarkson's database is shown in table 1 in the annexes.

All coloured empty cells in the table refers to data not available in Clarkson database

3.2. HATCH AREAS

Hatch areas are computed and tabulated in table 2 below:

Ship	Hatch Area (m^2)
Ship 1	4240.80
Ship 2	3800.52
Ship 3	4561.92
Ship 4	2271.74
Ship 5	2212.50
Ship 6	1716.34
Ship 7	1468.80
Ship 8	1754.22
Ship 9	1720.40
Ship 10	1935.36

Table 2: Hatch areas calculated from hatch dimensions

3.3 SIMULATION TO DETERMINE AVERAGE ELECTRICAL LOAD USING MONTE CARLO CRYSTAL BALL

In order to total auxiliary power rating of each ship, sum of individual power ratings *P*(kw) of the auxiliary engine are summed up and maximum electrical load, *Emax* (Kw) taking a load factor *f* of 75% is calculated using equation 1. According to [12], auxiliary diesel engines usually have MCR ranging 50-75%. Since electrical load and load factor are variable quantities depending on the electrical load demand, average load on the engine corresponding to ship's average electrical load is determined using equation 4.

Using total auxiliary power rating and engine load factor as decision variables and maximum electrical load as the forecast, average electrical load is obtained by Monte Carlo Crystal Ball simulation. Assumption cells contain values of available load range from 50-100% and permissible range of engine load factor 50-75%.

The result of simulation is given in Table 3 below

SHIP	total auxiliary rating P(kw)	load factor f	maximum electrical load Emax (KW)	Average Elecrical load(KW)
SHIP1	3762	0.75	2821.5	1404.63
Ship 2	1440	0.75	1080	541.04
Ship 3	5120	0.75	3840	1922.98
Ship 4	1650	0.75	1237.5	618.28
Ship 5	2400	0.75	1800	899.4
Ship 6	1470	0.75	1102.5	827.55
Ship 7	1331	0.75	998.25	500.14
Ship 8	1950	0.75	1462.5	727.75
Ship 9	1830	0.75	1372.5	611.33
Ship 10	1800	0.75	1350	674.99

Table 3: Average power determined by simulation

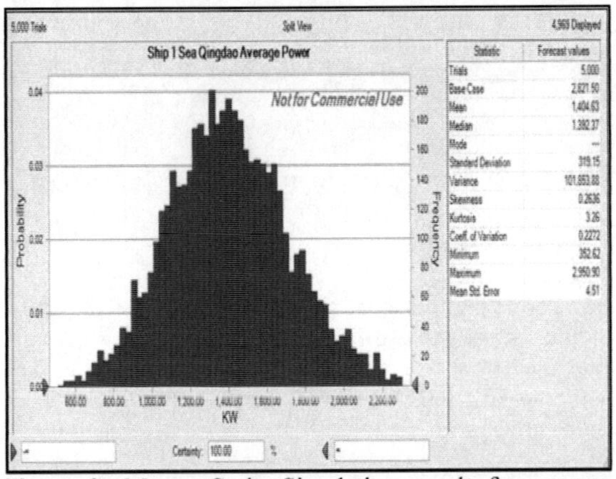

Figure 2: Monte Carlo Simulation result for average power

3.4. CALCULATION OF AVERAGE DAILY FUEL CONSUMPTION

Similarly, average daily fuel consumption is determined using equations (3) and (5). Eq (3) is used to first determine the maximum fuel consumption, using the maximum engine load or electrical load condition and specific fuel oil consumption of 200g/KWh for medium and high speed engines. Using eq. (5) to determine the average daily consumption, Monte Carlo Crystal Ball Simulation is used. In this case, maximum fuel consumption is used as the forecast, whilst decision variables (green cells) are the total auxiliary power rating and load factor on which fuel consumption depends. The same ranges of values for assumptions used in the average load determination are used.

The simulation values average daily fuel consumption (light green cells) obtained are shown in table 4 below:

Sf	200	g/kwh	24	hrs
	total auxiliary rating p (kw)	load factor f	Max Fuel Consumption of (tons)/day	Qav Average consumption/day (tons)
SHIP1	3762	0.75	13.5432	6.77
Ship 2	1440	0.75	5.184	2.59
Ship 3	5120	0.75	18.432	9.22
Ship 4	1650	0.75	5.94	2.97
Ship 5	2400	0.75	8.64	4.33
Ship 6	1470	0.75	5.292	2.65
Ship 7	1331	0.75	4.7916	2.38
Ship 8	1950	0.75	7.02	3.52
Ship 9	1830	0.75	6.588	2.94
Ship 10	1800	0.75	6.48	3.23

Table 4: Average fuel consumption determined by simulation

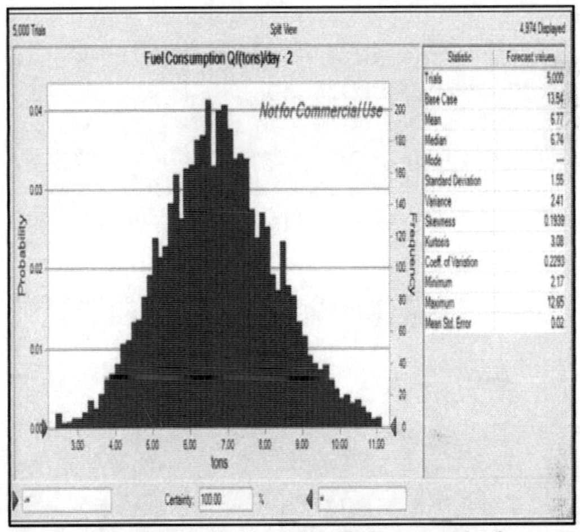

Figure 3: Monte Carlo Simulation graph for average daily fuel consumption

3.5. SOLAR POWER DETERMINATION

In determining the power generated by the PV system based on the available hatch area, specification of Trina solar 72 cells polycrystalline silicon module is used. Module efficiency 17.1% at 335w_p, and extreme dimensions 1984mm by 992mm, power generated by PV system using equation 6 is tabulated in table 5 below:

S/N	Hatch Area (m2)	PV power output (kw)
Ship 1	4240.8	368.48
Ship 2	3800.52	330.23
Ship 3	4561.92	396.39
Ship 4	2271.744	197.39
Ship 5	2212.5	192.24
Ship 6	1716.338	149.13
Ship 7	1468.8	127.62
Ship 8	1754.22	152.42
Ship 9	1720.4	149.49
Ship 10	1935.36	168.16

Table 5: PV system potential power produced (KW) and hatch area

3.6. ENGINE LOAD WITH PV SYSTEM PRODUCING POWER *(HYBRID CASE)*

With the PV system producing power during the day, diesel engine in the hybrid case therefore carry partial load. This is determined using Eq. 7. The result is shown in table 6 below:

S/N	Area (m2)	PV power output (kw)	Load on Aux. Engines Conv (KW)	Load on Aux Engines Hybrid(KW)
Ship 1	4240.8	368.48	1404.63	1036.15
Ship 2	3800.52	330.23	541.04	210.81
Ship 3	4561.92	396.39	1922.98	1526.59
Ship 4	2271.744	197.39	618.28	420.89
Ship 5	2212.5	192.24	899.40	707.16
Ship 6	1716.338	149.13	827.55	678.42
Ship 7	1468.8	127.62	500.14	372.52
Ship 8	1754.22	152.42	727.75	575.33
Ship 9	1720.4	149.49	611.33	461.84
Ship 10	1935.36	168.16	674.99	506.83

Table 6: Engine load for Conventional and Hybrid cases with PV system producing power

3.7. CALCULATION OF AVERAGE DAILY FUEL CONSUMPTION IN THE HYBRID CASE

Using Eq. (8), the average daily consumption is determined taking S_f as 200g/kwh. The difference between the two scenarios is the PV power equivalent of fuel consumption saved by the operation of the PV system. The result is given in table 7 below:

S/N	Area (m2)	PV power output (kw)	Load on Aux. Engines Conv (KW)	Load on Aux Engines Hybrid(KW)
Ship 1	4240.8	368.48	1404.63	1036.15
Ship 2	3800.52	330.23	541.04	210.81
Ship 3	4561.92	396.39	1922.98	1526.59
Ship 4	2271.744	197.39	618.28	420.89
Ship 5	2212.5	192.24	899.40	707.16
Ship 6	1716.338	149.13	827.55	678.42
Ship 7	1468.8	127.62	500.14	372.52
Ship 8	1754.22	152.42	727.75	575.33
Ship 9	1720.4	149.49	611.33	461.84
Ship 10	1935.36	168.16	674.99	506.83

Table 7: Average daily fuel consumption in Hybrid and Convention case

3.8 CALCULATION OF LUBRICATING OIL CONSUMPTION IN CONVENTIONAL AND HYBRID CASES

The lubricating oil consumption is determined using specific lubricating oil consumption S_L *(g/kwh)* of Wartsilla engines of similar power rating which is 0.6 [18] for engine power ranges between 525-735KW. The results are shown in table 8 below using equations 9 &10.

S/N	Load on Aux Engines Conv (KW)	Load on Aux Engines Hybrid(KW)	Annual Lube Cons Conv (tons)	Annual Lube cons. Hybrid (tons)
Ship 1	1404.63	1036.15	7.38	5.44
Ship 2	541.04	210.81	2.84	1.11
Ship 3	1922.98	1526.59	10.12	8.02
Ship 4	618.28	420.89	3.23	2.21
Ship 5	899.40	707.16	4.73	3.72
Ship 6	827.55	678.42	4.35	3.57
Ship 7	500.14	372.52	2.63	1.96
Ship 8	727.75	575.33	3.83	3.02
Ship 9	611.33	461.84	3.21	2.43
Ship 10	674.99	506.83	3.55	2.66

Table 8: Annual lube oil consumption for Conventional and Hybrid Cases

3.9. COST ELEMENTS

The cost elements and total energy produced by the system calculated over one year period are the inputs to the computation of the Levelised Cost of Energy. Cost elements for conventional and hybrid cases are similar with additional cost elements of PV system for the hybrid case. Rotterdam bunker prices of February, 2017 for Heavy Fuel Oil are used and price piece for a select lube oil supplier based on the recommendation of major auxiliary engine builders under the brand name AURELIA TI 4055[18]. For cost relating to Capex, prices available in recent literatures for auxiliary diesel engines [19] and PV modules are used. Table 9 below is the result of the cost elements.

S/N	FUEL COSTS		LUBE OIL COSTS		CAPEX		OPEX	
	Annual Fuel Cost Conv ($)	Annual Fuel Cost Hybrid ($)	Lub Cost Conv ($)	Lub Cost Hybrid ($)	Capex Aux DE ($)	Capex Solar PV ($)	Opex Conv ($)	Opex Hybrid($)
Ship 1	747492.6	649848	18518.57	13660.49	940500	191611.8	766011.2	663508.5
Ship 2	285968.4	198461.2	7133.04	2779.33	360000	171718.6	293101.4	201240.6
Ship 3	1018003	912964.8	25352.47	20126.53	1280000	206120.9	1043356	933091.4
Ship 4	327925.1	275618.1	8151.37	5548.96	412500	102644	336076.5	281167.1
Ship 5	478086.1	427143.2	11857.64	9323.10	600000	99967.23	489943.8	436466.3
Ship 6	292593.1	253074.4	10910.37	8944.21	367500	77549.17	303503.5	262018.6
Ship 7	262781.8	228962.6	6593.82	4911.23	332750	66364.68	269375.6	233873.8
Ship 8	388652	348261	9594.62	7585.06	487500	79260.79	398246.6	355846.1
Ship 9	324612.8	285000.5	8059.74	6088.93	457500	77732.71	332672.5	291089.4
Ship 10	356632.4	312070.6	8899.03	6681.97	450000	87445.23	365531.4	318752.6

Table 9: Cost breakdown for Conventional and Hybrid Bunker cost US$/ton=302.5; Lubricating oil price US$/barrel=480

3.10. ENERGY PRODUCED

The energy produced in both cases is considered to be equal due to the fact that the PV system in the hybrid case is an auxiliary to the main power pack. Table 10 below gives the energy produced by auxiliary engine during the period of 365days as well as the energy that the PV system would have produced during the same period for generating during the day.

S/N	PV power output (kw)	Load on Aux Engines Conv (KW)	Load on Aux Engines Hybrid(KW)	PV Energy supplied (kwh)	Total Energy (kwh)
Ship 1	368.48	1404.63	1036.15	1613960.56	12304559
Ship 2	330.23	541.04	210.81	1446399.12	4739510
Ship 3	396.39	1922.98	1526.59	1736172.18	16845305
Ship 4	197.39	618.28	420.89	864578.67	5416133
Ship 5	192.24	899.40	707.16	842031.63	7878744
Ship 6	149.13	827.55	678.42	653202.66	7249338
Ship 7	127.62	500.14	372.52	558994.83	4381226
Ship 8	152.42	727.75	575.33	667619.76	6375090
Ship 9	149.49	611.33	461.84	654748.57	5355251
Ship 10	168.16	674.99	506.83	736557.89	5912912

Table 10: Total energy produced by the auxiliary diesel engine and PV system over one year duration

3.11. ANNUALISED TOTAL COST

In the determination of the annualised total cost, the length of life of ship and the PV systems are taken as 30yrs [9] and the discounted rate of 8% usually taken for investment in maritime infrastructures [7] is used for the computation of the annualised factor. Table 11 below gives the annualised TC.

S/N	Total Cost Conv($)	Total Cost Hybrid($)	Annualised Total Cost Conv($)	Annualised Total Cost Hybrid($)
Ship 1	1706511	1795620	152220.8	160169.3
Ship 2	653101.4	732959.2	58256.65	65379.96
Ship 3	2323356	2419212	207243.3	215793.7
Ship 4	748576.5	796311.1	66773.02	71030.95
Ship 5	1089944	1136434	97222.98	101369.9
Ship 6	671003.5	707067.8	59853.51	63070.44
Ship 7	602125.6	632988.5	53709.6	56462.57
Ship 8	885746.6	922606.9	79008.6	82296.53
Ship 9	790172.5	826322.1	70483.39	73707.93
Ship 10	815531.4	856197.8	72745.4	76372.85

Table 11: Annualized total cost for Conventional and Hybrid CRF =0.0892 calculated from Eq. 12

4. DISCUSSION OF RESULTS AND SCENARIO ANALYSIS

Discussion of results of this study will be dealt with fewer than two broad areas as follows:

Current Findings relating to:

a. LCOE comparison under existing market condition with respect to bunker cost, PV module price, and efficiency and Capex of auxiliary diesel engines provided in the literature

b. CO_2 emission reduction potentials under existing situation.

Future Scenarios associated with:

a. Effect of changes in module price and efficiency on LCOE

b. Effect of changes in bunker costs on LCOE using Rotterdam price for HFO 380 cST

4.1 CURRENT FINDINGS

a. LCOE comparison between conventional and hybrid cases for which two situations exists with hybrid case.

These are the hybrid retrofit (existing) vessel and hybrid (new). In the retrofit case, total energy is equivalent to energy produced by existing auxiliary diesel engines, whilst for new vessel case, auxiliary engines are hybridised with PV ay the design stage. In this case total system energy is the sum of energy produced by auxiliary engines and PV system. Under current market prices of bunker, module efficiency and module price of 0.52$/KW, LCOE comparison is shown in figure 4 below:

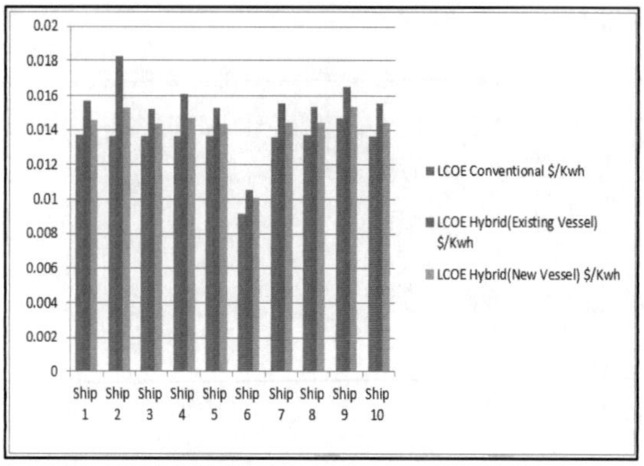

Figure 4: LCOE comparison-current findings

As LCOE is a measure of cost energy over an assumed life time of a generating plant, the smaller the value the more cost effective it is to operate a particular technology for power (electricity) generation.

In figure 4 above the comparison shows a lower LCOE value for the conventional scenario than the hybrid case in which the PV system is retrofitted to the existing vessels. This is to be expected since in this case the total energy of the system is considered to be equal to the energy produced by the existing auxiliary diesel engine(s) whilst the energy produced by the PV is not accounted in the determination of the LCOE whereas its capital cost has been factored. The reason for this is that, the PV system is regarded as an auxiliary to the main power pack. However, in the case where the PV system is considered at the design of a new vessel and its energy production is also considered as forming part of total energy of the hybrid system will result to lower LCOE when compared to the conventional case where only auxiliary diesel engine is installed.

Cost Savings under current findings:

Even where the LCOE in the hybrid case is higher than the conventional case due to the engine running at part load, there is savings in fuel and lubricating oil consumption of about 10-31% and 18-61% respectively over one year period. Figure 5 below shows the savings for various ships. The percentage savings are proportional to the power produced by the PV system on each ship under the assumed standard testing conditions of the polycrystalline silicon module with 17.1% efficiency.

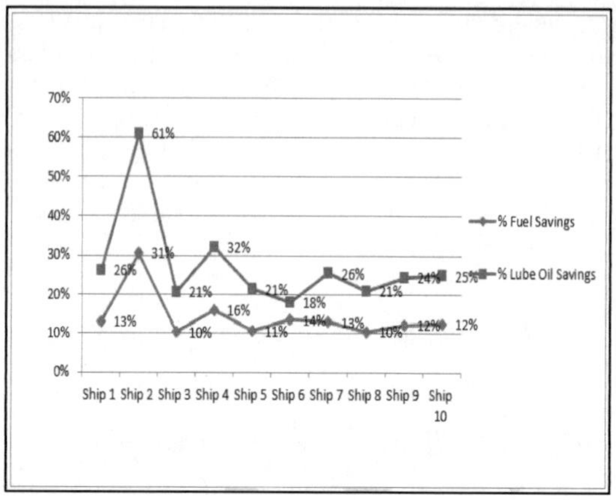

Figure 5: Current percentage savings in fuel and lube oil consumption in the Hybrid (Existing) vessel

b. Effects of current findings on possible co_2 emissions reduction:

The amount of carbon dioxide emitted by the auxiliary diesel engine is proportional to the amount of fuel consumed. In the third IMO GHG study, the total annual CO_2 emissions by bulk carriers is calculated using the weighted average based on the deadweight bracket, number of days at sea and average installed engine power.

S/N	Dwt	Amount of Annual CO2 Reduction (tonnes) Due to PV Application	% Reduction of Annual Total CO_2 Emissions
Ship 1	397,539	1003.8835	2.7%
Ship 2	206,204	899.66025	2.4%
Ship 3	105,708	1079.8991	3.7%
Ship 4	99,047	537.76793	2.7%
Ship 5	91,439	523.74367	2.6%
Ship 6	82,165	406.29205	2.0%
Ship 7	68,788	347.69478	1.7%
Ship 8	61,965	115.25949	2.1%
Ship 9	50,913	407.25361	2.8%
Ship 10	41,536	458.13901	3.2%

Table 12: Potential annual CO2 emission reduction due to PV application on bulk carriers

Using the carbon factor of 3,114 kg/ton of fuel burned (IMO GHG Study, 2014), the table above shows the annual CO_2 emissions reduction potentials due to PV application on existing bulk carrier ships discussed in this research. The percentages are expressed in terms of total annual CO_2 emissions from the ships based on the information provided on bulk carriers in the IMO Third GHG Study. Graphical representation of the reduction in CO_2 is shown below.

The reduction potential is proportional to the power produced by the solar PV under the operating conditions. However, the percentage value is relative to the total annual CO_2 emissions from the ship. This depends on total installed power, deadweight, and the availability of boiler and shaft generators. The total emission also depends on the number of sea days as the consumption is different when the main is operating and when the vessel is in port. Therefore for vessels of similar size and installed power, there is more CO_2 reduction potential when PV is installed than the conventional case power is supplied by only auxiliary diesel engine.

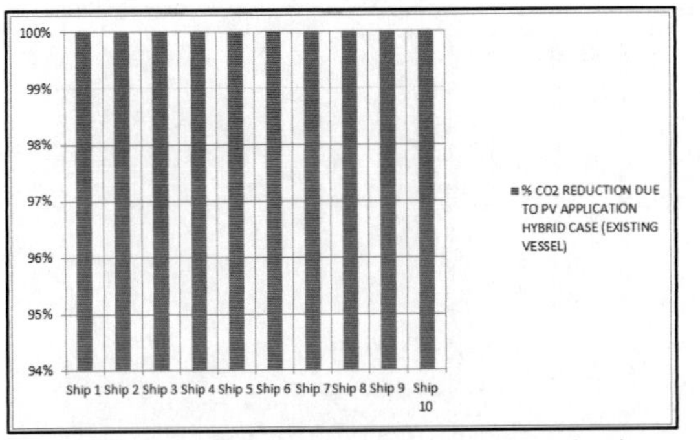

Figure 6: Potential annual CO2 emission reduction

4.2 FUTURE SCENARIO DEMONSTRATION OF SOLAR PV POTENTIAL

a. Effect of changes in module price and efficiency on LCOE.

Demonstration of the future potentials of the D/E-PV hybrid on the bulk carriers covered in this study will utilize the changes in fuel price which is a major variable in shipping as it constitute 20-60% [2] of ship's operating cost ad changes in PV module price and efficiency. Maintaining auxiliary diesel engine Capex at 300\$/kW and using module price of 0.28\$/kw and efficiency of 23% as predicted for 2025 [10], effect these changes on LCOE is indicated in the figure 7 below.

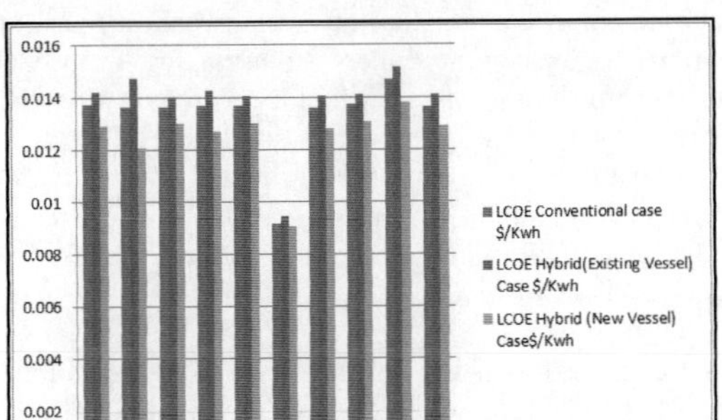

Figure 7: Future scenario LCOE comparison

This indicates a lower LCOE for the hybrid (New Vessel) case than the conventional case and where the conventional case has lower LCOE than hybrid (existing vessel) case, there is a bigger fuel and lube oil savings as depicted in the figure below.

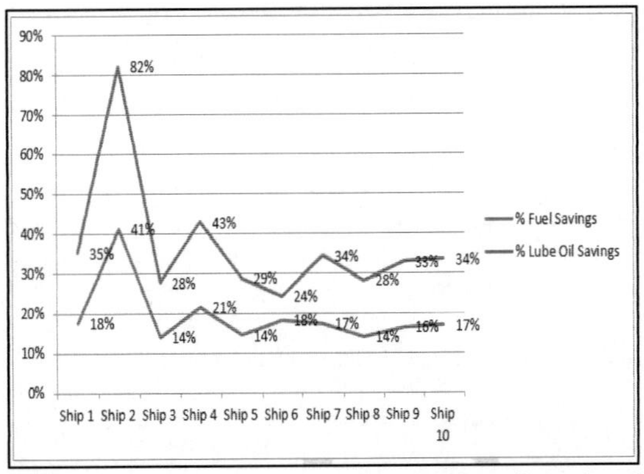

Figure 8: Future scenario fuel and lube oil saving in Hybrid (Existing)

b. Effect of changes in bunker costs on LCOE using Rotterdam price for HFO 380 cST.

Maintaining the module efficiency and price as predicted for 2025, and Considering the lowest and highest Rotterdam bunker prices of 144\$/ton and 640\$/ton occurring between 2004 and 2017, effect bunker changes on LCOE is given in figure 9 below:

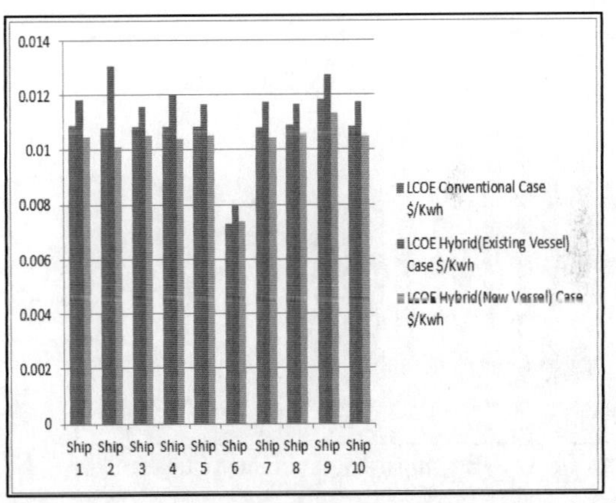

Figure 9: LCOE comparison at 2016 Rotterdam bunker price for 380Cst HFO @144\$ per TON

It is evident from figure 28 above that when bunker prices are at its lowest of 144\$/ton, the LCOE for the hybrid (Existing Vessel) case is lower when compared to the current scenario bunker prices of 302.5\$/ton.

The annual savings from fuel consumption due to the PV application valued from \$21,500-\$67,200 is less when compared to the savings at the bunker price of 302.5\$/ton which is \$33,000-\$105,000. This is shown in figure 10 below.

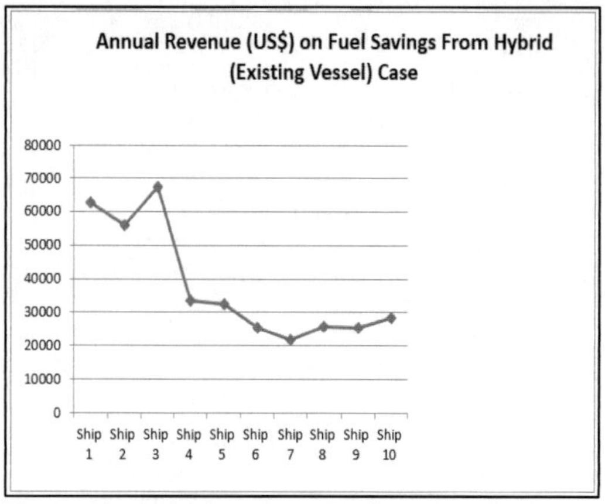

Figure 10: Annual revenue (US$) on fuel savings from hybrid case (existing vessels) at lowest bunker price

When the highest average bunker price of approximately 640$/ton is used the values of LCOEs are generally higher than at lower bunker price. However, the hybrid case for both existing and new vessels has lower LCOE than the conventional case. This is an important discovery as it indicates that when bunker prices soar, it is more cost effective to have a hybridised system. This is shown in figure 11 below:

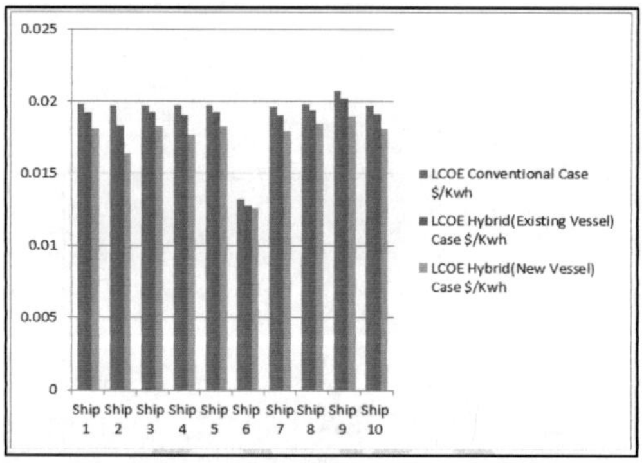

Figure 11: LCOE comparison at highest bunker

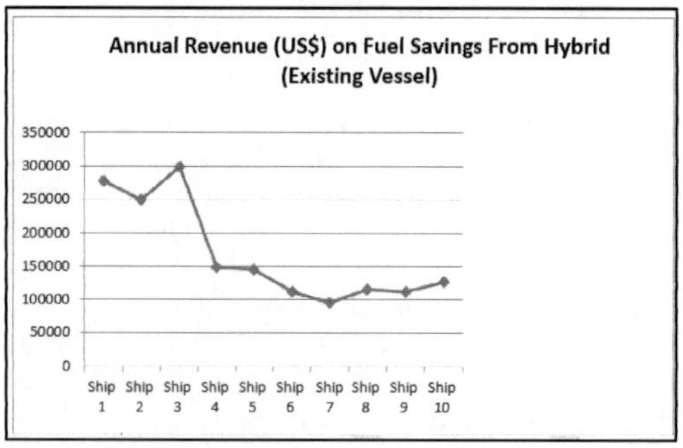

Figure 12: Revenue on fuel savings at the extreme bunkerpriceof$640/Ton

The fuel savings in terms of revenue at the extreme bunker price of $640/ton is $112,000-$300,000 for the model case with PV complementing as shown in figure 12 above.

c. Effects of future scenario on carbon dioxide co_2 emissions.

When a module efficiency of 23% is considered, the possible CO_2 emissions reduction is shown in the figure 13 below:

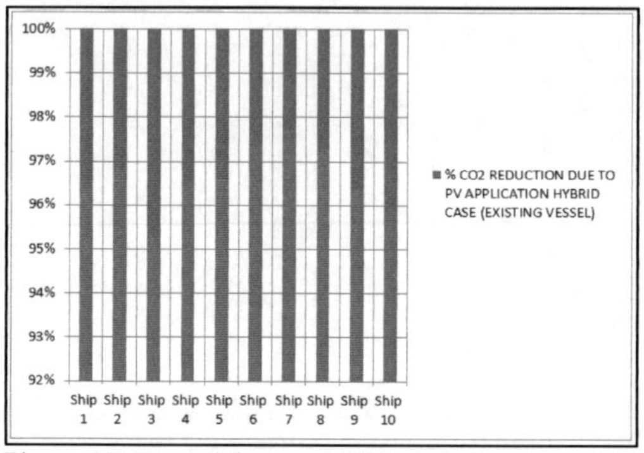

Figure 13: Potential annual CO2 emissions reduction due to PV application on the bulk carriers

As power produced by PV system depends on the module efficiency, there is reduction in fuel consumption when more power is produced by the PV system and consequent reduction in emissions. There is higher percentage reduction in CO_2 emissions under future scenarios compared to current scenario. The complete work of this research can be found in the first author's MSc dissertation conducted at the World Maritime University [21].

5. CONCLUSIONS

The unprecedented growth in the world seaborne trade over the last decade due to population growth and industrialisation has resulted in the increase in maritime transport. A consequential increase in the amount of CO_2 emitted from the use of fossil fuel by merchant ships has been recorded in the IMO GHG Studies. Bulk Carrier fleet constitute a significant number of the world merchant fleet and currently has 43.1% share by deadweight of the world deadweight tonnage according to UNCTAD 2016 report.

Whilst every effort is being made by IMO to respond to the global call for CO_2 reduction in the international shipping through technical and operational measures, many technological innovations such as LNG, wind, methanol and fuel cells have been researched into and some of which are now finding applications or being considered for merchant ships, the case of Solar PV has received little attention despite its many applications on land.

Consequently, this research investigated into the potential of solar PV on dry bulk carriers in auxiliary power capacity whose open space on the hatch covers are never utilised throughout the ship's lifetime. In the absence of vital data regarding annual fuel consumption of the selected ships, this research has developed a methodology with high degree of confidence for arriving at the values used in Chapter IV and applied LCOE concept to determine the PV potentials.

The research has discovered significant outcomes, which can be summarised as follows:

1. Under current market price scenarios, LCOE for the conventional case is lower compared to the hybrid case due to high Capex and relatively low module efficiency. However even under the current price scenarios, with PV application there is fuel savings of between 10-31% compared to conventional over one year period. There is more to gain in the long run from the environmental perspectives due to the CO_2 reduction potentials from reduced fuel consumption.

2. Future scenario prediction where bunker price, module efficiency and module price are variables, the hybrid LCOE responds more favourably than the LCOE for conventional case even with lower bunker price scenario. With higher bunker cost as was the case in 2012, the hybrids comparatively have much lower LCOE than the conventional case.

3. Hybrid LCOE is sensitive to variation in module price and efficiency. With the speed of advancement in PV module technology, and the upping of the scale of module production, higher efficiency and low module prices are predicted. Therefore, there is more to gain in the application of technology on bulkers as LCOE will be lower when compared to conventional scenario of the same installed power. Additionally, with higher module efficiency there is more power generated by PV assuming STC prevails and consequently less fuel consumed by the hybridised diesel engine. This implies that there is a win-win situation in the utilization of Solar PV due to less emission of CO_2 and low operating cost.

4. A potential CO_2 emissions reduction of 2-5% annually of the total annual CO_2 emissions from the selected dry bulk carrier ships is possible with the application of a renewable energy option such as solar PV to assist auxiliary diesel engines in auxiliary power generation capacity.

5. As this research is representative of the overall bulk carrier fleet of 11,083 ships, if it is assumed that the PV technology is applicable to only 10% of the fleet, then an average of nearly 3% CO_2 emissions reduction of total annual CO_2 emissions from nearly 1110 bulk carrier ships is possible.

This research has revealed large potentials of Solar PV applications on the selected dry bulk carriers using a combination of methodologies, taking into consideration the key assumptions and limitations stated.

Further research is therefore recommended with the objective of reducing the assumptions and limitations imposed especially in the following areas:

- Investigate the potential of Solar PV on a particular sea route using time and location for solar irradiance. This takes into account the effects environmental variabilities on the solar irradiance;

- Increase the number of selected ships for each size category in order to depict a possible disparity in trend between categories.

6. REFERENCES

1. Caspian Publishing. Energy. *Professional engineering*, (31), 1–4. (2013)

2. Corbett, J. J., & Winebrake, J. *The impacts of globalisation on international maritime transport activity*. Energy and Environmental Research Associates, the United States, (November), 31. (2008).

3. Dalgic, Y. *Development of offshore wind operational expenditure model and investigation of optimum operation and maintenance fleet*. University of Strathclyde. (2015).

4. Fuentealba, E., Ferrada, P., Araya, F., Marzo, A., Parrado, C., & Portillo, C. *Photovoltaic performance and LCoE comparison at the coastal zone of the Atacama Desert,* Chile. Energy Conversion and Management, 95, 181–186. https://doi.org/10.1016/j.enconman.2015.02.036. (2015).

5. Green Futures. *Sun Powers the Dream*. Green Futures, pp. 1–2. Retrieved from www.greenfutures.org.uk. (2009)

6. IMO. *Third IMO GHG Study, 2014*. London. Retrieved from http://www.imo.org/en/OurWork/Environment/PollutionPrevention/AirPollution/Pages/Greenhouse-Gas-Studies-2014.aspx. . (2014).

7. IMO. *Studies on the Feasibility and use of LNG as a Fuel for Shipping*. Air Pollution and Energy

Efficiency Study Series (pp. 66–67). London: IMO. Retrieved from www.imo.org. (2016).

8. International Energy Agency IEA. *Solar Energy Perspectives*. Solar Energy Perspectives. https://doi.org/10.1787/9789264124585-en. (2011).

9. International Energy Agency IEA. *Solar Photovoltaic Energy*. Technology Roadmap, 12–25. https://doi.org/10.1007/SpringerReference_7300. (2014)

10. IRENA. *The Summer of Solar*. IRENA Quarterly, (3rd Quarter), 1–8. Retrieved from www.irena.org/publications. (2016).

11. Lan, H., Wen, S., Hong, Y. Y., Yu, D. C., & Zhang, L. *Optimal sizing of hybrid PV/diesel/battery in ship power system*. Applied Energy, *158*, 26–34. https://doi.org/10.1016/j.apenergy.2015.08.031. (2015).

12. Molland, Anthony, F. *A Maritime Engineering Reference Book: A Guide to Ship Design, Construction, and Operation*. Amsterdam: Butterworth-Heinemann. (2008).

13. Salem, A. A., & Seddiek, I. S. *Techno-Economic Approach to Solar Energy Systems Onboard Marine Vehicles*. Polish Maritime Research, 23(3), 64–71. https://doi.org/10.1515/pomr-2016-0033. (2016).

14. Seijo, G., & Formoso, F. *Alternative Sources of Energy in Shipping*, (10), 435–448. https://doi.org/10.1017/S0373463310000111. (2017).

15. Smith, T. W. P., Jalkanen, J. P., Anderson, B. A., Corbett, J. J., Faber, J., Hanayama, S., … Hoen, M., A. *Third IMO Greenhouse Gas Study 2014*. International Maritime Organization (IMO*)*. https://doi.org/10.1007/s10584-013-0912-3. (2014).

16. Teeter, J. L., & Cleary, S. A. *Decentralized oceans: Sail-solar shipping for sustainable development in SIDS*. Natural Resources Forum, 38(3), 182–192. https://doi.org/10.1111/1477-8947.12048. (2014).

17. UNCTAD. *Transport Review*. (2016).

18. WÄRTSILÄ. *Wärtsilä Auxpac Product Guide*. (2017).

19. Welaya, Y. M. A., Mosleh, M., & Ammar, N. R. *Thermodynamic analysis of a combined gas turbine power plant with a solid oxide fuel cell for marine applications*. International Journal of Naval Architecture and Ocean Engineering, 5(4), 529–545. https://doi.org/10.2478/IJNAOE-2013-0151. (2013).

20. Yoshida, S., Ueno, S., Kataoka, N., Takakura, H., & Minemoto, T. *Estimation of global tilted irradiance and output energy using meteorological data and performance of photovoltaic modules*. Solar Energy, 93, 90–99. https://doi.org/10.1016/j.solener.2013.04.001.(2013).

21. *Wandifa Saidyleigh. Investigation of Auxiliary Power Potentials of Solar Photovoltaic Applications on Dry Bulk Carrier Ships (Master Dissertation). Malmo, Sweden: World Maritime University. (2017).*

7. AUTHORS BIOGRAPHY

Wandifa Saidyleigh is a marine engineer and holds the position of Marine Surveyor in the Gambia Maritime Administration. He has an MSc in the maritime affairs specializing in maritime energy management with a special focus on renewable energy use onboard ships.

Dr. Aykut I Ölçer is a naval architect and marine engineer holding the position of Nippon Foundation Professorial Chair in Marine Technology and Innovation at the World Maritime University (WMU), Malmö, Sweden. He currently leads the WMU Maritime Energy Research Group (http://wmu.se/research) as well as serves as the Head of Maritime Energy Management Specialisation at WMU. At the same time, he is the Editor in Chief of WMU Journal of Maritime Affairs (JOMA) and Book Series.

Dr. Raphael Baumler holds the position of Associate Professor at the WMU, in Malmo-Sweden. His expertise revolves around the extended Maritime Safety and Environmental Administration (MSEA) domain focusing on the impact of the vessel's socioeconomical environment on safety and environment. Primarily educated as a dual officer, he has worked on various types of vessels as Captain. Dr. Baumler spent 20 years in a seafaring career.

8. ANNEXES

	SHIP1	SHIP2	SHIP3	SHIP4	SHIP6	SHIP7	SHIP8	SHIP9	SHIP10	SHIP5
Length Overall (LOA)	361	299.94	234.93	249.94	228.99	225	225	189.99	185	235
Breadth Moulded	65	50	43	43	32.26	32.2	32.26	32.26	31	43
Gross Tonnage (GT)	199959	104,732	73,427	55,284	43,022	37,519	38,573	28,647	27,781	50,248
Deadweight(Dwt)	397539	206,204	105,708	99,047	82,165	68,788	64,965	50,913	41,536	91,439
Year of Build	2012	2005	2001	2005	2013	1990	2014	2001	2017	2000
Main Engine Make	MAN B&W 7S80ME-C8	Man B&W 6S70MC-C7	Mitsubishi 8 UEC60 LSII	Man B&W-6S60MC-C7	Man B&W-6S60MC7	Man B&W-6S60MC MK3	Man B&W-6S60ME-C8	Man B&W-6S50MC-C	Man B&W-5G50ME-B9	Mitsubishi-6UEC60LSII
Main Engine SFOC (g/KWh)										
Main Engine Power Rating (kw)	29260	18,680	15298	10400	9710	11256	9660	9466	4900	11915
Auxilairy Engines(Generator Sets) Make	MAN Turbo 6L 21/31	Yanmar 6N21AL-DV	Daihatsu 5DK-26	Yanmar - 6N18AL-UV	Daihatsu-5DC-17Ae	Man Diesel Turbo-5L23/30	Man Diesel Turbo-5L23/30	Daihatsu-5DK-20	Man Diesel Turbo-5L23/30	Daihatsu-5DK-20
Auxiliary Engines Power Rating (total) (KW)	3762	1440	5120	1650	1470	1331	1950	1830	1800	2400
Number of Auxiliary Engines	3	3	4	3	3	2	3	3	3	3
Auxilairy Engine SFOC(g/KWh)										
Fuel Type used by Auxilairy Engines	HFO	HFO	HFO	HFO	HFO	HFO	HFO	HFO	HFO	HFO
A.E Daily Fuel Consumption(tons)										
Auxiliary Engines Total Annual Lube Oil Consumption										
Auxilairy Engines Lube Oil Type and Grade										
Ship Minimum Electrical Load										
Ship Maximum Electrical Load										
Ship's Number of Operating Days per Year										
Route Most Used by Ship	Long Haul Brazil-China	Long Haul Brazil-China	Long Haul Brazil-China	Long Haul Brazil-China	Long Haul Brazil-China	Long Haul Brazil-China	Short Haul	Short Haul	Short Haul	Long Haul-Brazil-China
Hatch Dimensions (L Xb)										
Hatch 1	28.50*20.46	16.20*20.40	28.80*26.40	19.20*15.66	15.37*15.00	14.40*12.00	15.57*12.00	17.85*17.60	19.20*14.40	21.00*18.30
Hatch 2	28.50*22.32	16.20*23.80	28.80*26.40	20.88*17.60	17.30*15.00	14.40*15.00	17.30*15.10	18.70*17.60	21.60*19.20	18.00*16.40
Hatch 3	28.50*22.32	16.20*23.80	28.80*26.40	20.88*19.20	17.30*15.00	14.40*15.00	17.30*15.10	20.40*17.60	21.60*19.20	21.00*14.60
Hatch 4	28.50*22.32	16.20*23.80	28.80*26.40	20.88*19.20	17.30*15.00	14.40*15.00	17.30*15.10	20.40*17.60	21.60*19.20	21.00*14.60
Hatch 5	28.50*22.32	16.20*23.80	28.80*26.40	20.88*19.20	17.30*15.00	14.40*15.00	17.30*15.10	20.40*17.60	21.60*19.20	21.00*14.60

Hatch 6	28.50*2 2.32	16.20*2 3.80	28.80*2 6.40	20.88*1 9.20	17.30*1 5.00	14.40*1 5.00	17.30*1 5.10			21.00*1 4.60
Hatch 7	28.50*1 6.74	16.20*2 3.80			14.71*1 2.80	14.40*1 5.00	17.30*1 5.10			21.00*1 4.60
Hatch 8		16.20*2 3.80								
Hatch 9		16.20*2 3.80								

Table 1: Raw data on dry bulk carriers (CLARKSON, 2017)

MODERN ROTOR SAIL TECHNOLOGY HELPS SHIPS SAVE FUEL AND REDUCE EMISSIONS - PERFORMANCE AND EXPERIENCES FROM RECENT INSTALLATIONS

V Paakkari, Norsepower Oy Ltd, Finland

SUMMARY

With an increasing need to reduce the emissions related to global shipping, means of emission free propulsion are required. Use of auxiliary wind propulsion is a natural choice as wind is free, globally available, and it is provided "on-site" instead of having to be transported with the ship from the port. Norsepower has modernised the Flettner rotor concept to obtain a highly efficient and robust solution to utilize the propulsive potential of the wind. Norsepower Rotor Sails harness the wind to provide a clean source of propulsion for various ship types. This paper addresses the need and means to provide the shipping market with verified information on the obtained savings with Rotor Sails. The paper also reviews the experiences of recent installations and finally provides an outlook of the global wind propulsion potential with respect to the busiest shipping routes.

1. INTRODUCTION

1.1 REDUCING EMISSIONS ON GLOBAL SHIPPING

The global shipping industry is challenged with increasing demands to reduce the environmental impact of shipping operations. Pressure to build and operate more environmentally sustainable ships was defined in more concrete terms by the IMO's recent decision to reduce the GHG emissions from international shipping by at least 50% by 2050.

Currently, the global shipping fleet is powered almost entirely by fossil fuels and the proven solutions available for achieving the 50% emission reduction are limited. New innovations and technologies are developed, which will provide the required solutions, however old innovations can also be utilised and modernised to deliver the necessary fuel efficiency improvements and emission reductions.

1.2 ROTOR SAIL TECHNOLOGY

Rotor Sails are a modernized version of Flettner rotors, almost a hundred-year-old invention of German engineer Anton Flettner. The concept is based on a principle called Magnus effect, which has been known in fluid dynamics even longer. The basic principle is, that a rotating object induces pressure difference between its sides when it is exposed to a flow(wind). To understand the origin of the pressure difference, one must study the direction of motions taking place. Figure 1 shows a schematic situation, where a round, rotating object is exposed to wind. When the object is rotating, a small boundary layer is formed around it. This is the layer of air that can be said to follow the surface of the object. Then the direction of the air flow in the boundary layer is same as the surface velocity, meaning it's pointing to different directions on different sides of the object. As the direction of the airflow due to wind is uniform throughout the scene, the relative direction of these two "separate" flows is different.

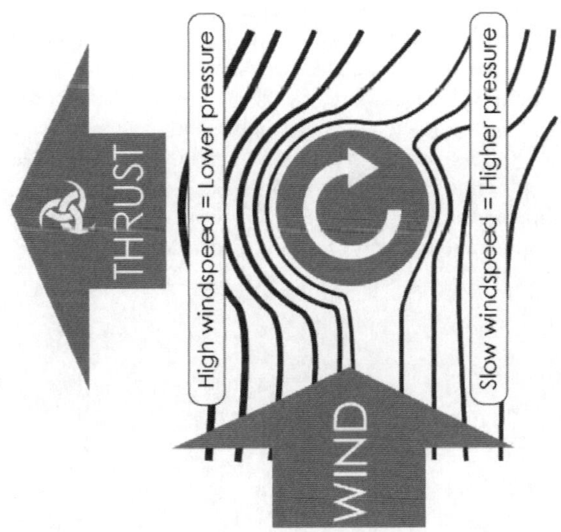

Figure 1: Schematic of the Magnus effect. Spinning round object is exposed to wind depicted with a red arrow.

In the convention of the Figure 1, the wind on the right-hand-side of the object encounters boundary layer flow with opposite direction and on the left-hand-side the flows are in the same direction. This means, that on the right-hand-side the rotational movement of the object resists, slows down, the wind flow and on the left-hand-side it enhances, speeds up, the wind flow. This change in wind speeds results in pressure difference according to so called Bernoulli principle; the increased speed leads to lower pressure, and decreased speed to higher pressure. The pressure difference between the sides results in a force vector, which is perpendicular to the wind (depicted with red "Thrust" arrow in figure). Rotor Sails are designed to transfer this force into the ship's hull to provide it with an additional thrust force, reducing the loading on main propeller. This then yields savings in fuel consumption of the vessel.

Norsepower has designed three standardised Rotor Sail models with heights of 18, 24, and 30 meters and corresponding rotor diameters of 3, 4 and 5 meters. A Rotor Sail consists of two main components:

- A rotating cylinder (the rotor), made of a laminated glass fibre and carbon fibre composite.
- A steel support tower inside the rotor extending up to circa two thirds of the rotor height.

The Rotor Sails are installed on vessel-tailored foundations, which are welded on to the deck at a shipyard. The Rotor Sails are installed on the foundations with a bolt connection. When the installation of the foundations has been completed, the Rotor Sails can be lifted on the vessel and connected to the foundations. To maintain the rotational movement, Rotor Sails are equipped with a variable speed drive with an electric motor, which is powered by the ships low-voltage power grid. The operation of Rotor Sails is fully automatic with respect to wind conditions and ships operation. The ship's crew can monitor the operation via a control panel on the bridge. In Figure 2, there is an example configuration of six Rotor Sails on a tanker vessel.

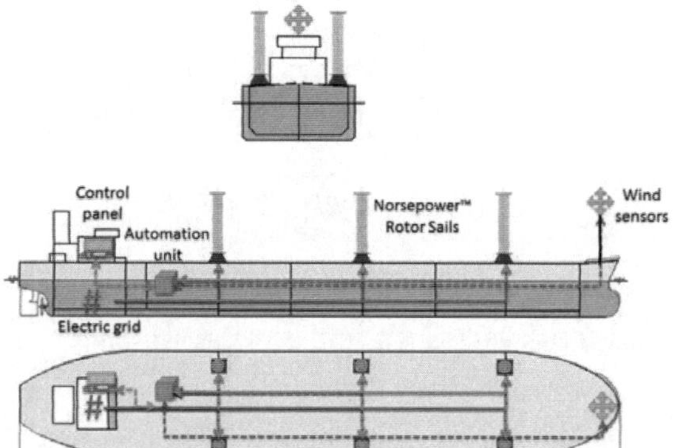

Figure 2: The Rotor Sail foundations are integrated to the ship's steel structure by welding. Cabling from the low voltage switchboard provides the required auxiliary power for the Rotor Sails and automation cabling connects the weather station sensors, Rotor Sail(s), and the control panel which is located on the bridge.

2. METHODS

2.1 INITIAL ANALYSIS

Rotor Sails save fuel by providing an auxiliary thrust force for the ship. To understand the economic and ecological profitability of the solution, the crucial metrics is the thrust force produced in certain wind conditions. There are multiple approaches to the problem from simple lifting-line considerations and more complex CFD calculations all the way to full-scale measurements.

In certain wind conditions, the performance of traditional sails even in the most modern configurations depends mostly on the angle of attack of the sail, which can be adjusted so that the maximum forward thrust is obtained. In comparison, the angle of attack for Rotor Sail is always dependent solely on the apparent wind direction. However, the performance can be influenced via controlling the rotational speed of the Rotor Sail. Then, before any performance tests can be done, one needs a model of the Rotor Sail performance in terms of rotational speed and apparent wind speed. The early performance model of Norsepower Rotor Sails was based on academic studies and computational simulations. From these, it was possible not just to estimate performance and potential propulsion power savings in different conditions, but to also build a model to optimize the Rotor Sail operation.

2.2 FULL-SCALE MEASUREMENTS

The computational and theoretical studies provide an excellent starting point to estimate Rotor Sail performance, but they should always be validated against real life measurements. At the time of early days of Norsepower, there was a very limited amount of publicly available full-scale test data of Flettner rotors. As in full-scale the Reynolds number effects may have influence and the inflow conditions are typically unsteady, it was essential to confirm the performance model against measured full-scale tests. The testing of the first prototype was divided into two stages. In the first stage, the Rotor Sail was tested on land. On the land test-site, it was also possible to calibrate the measurement system, as the Rotor Sail itself was rigidly installed on the ground.

Figure 3: Calibration of the measurement system

In the second stage, a sea-trial was conducted. During this trial NAPA, a leading maritime data analysis, software, and services provider was responsible for providing third party measurements. As on real-life conditions the ship and the Rotor Sail encounter many acceleration components and forces (wind, waves, …), it is not possibleto study simply the forces acting on the Rotor Sail in order to estimate the performance of the system. More feasible and straight forward way is to observe the propulsion power. Propulsion power requirement is a multivariable problem by nature, so it is important to rule out unnecessary variables. The first step towards this was to reduce the effect of environmental variables by collecting data over long time (10 months) with and without Rotor Sails operating. In practice, this was done by enabling / disabling the Rotor Sail for a certain time interval. Then the data, including thousands of data points, was analyzed by NAPA and as a result, two speed-power curves could be obtained. The other representing the vessel with Rotor Sail enabled, and the other one with Rotor Sail disabled.

3. ROTOR SAIL PERFOMANCE

3.1 PILOT PROJECT

A vital part of developing new technological solutions for marine applications is extensive full-scale testing and operation in the actual operating environment. Norsepower was very fortunate to have Bore as the first shipowner, who was prepared to install a Rotor Sail unit onboard the ro-ro vessel *M/V Estraden*. The first 18 m tall and 3 m diameter Rotor Sail was installed in the autumn of 2014. After less than a year in operation, Bore decided to install a second Rotor Sail unit to further improve the fuel saving performance and increase the benefits of wind-assisted propulsion.

The average fuel savings achieved with two Rotor Sails in the North Sea traffic has been approximately 6.1% – equalling circa 400 metric tonnes of fuel per year in reduced fuel consumption and more than $280,000 per year in reduced fuel costs (based upon Rotterdam October 2018 MGO price). This saving has been measured and independently verified by NAPA.

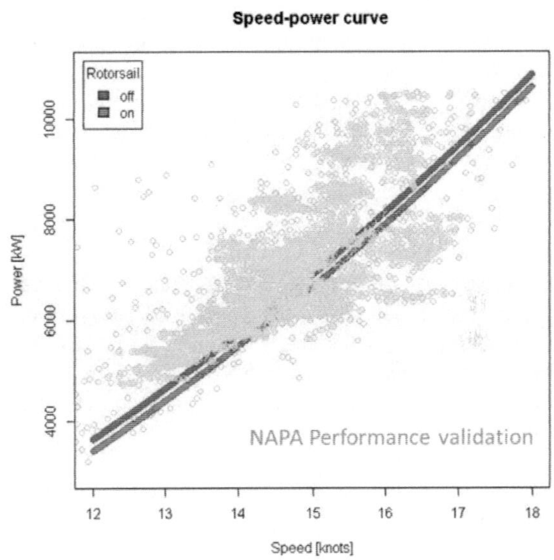

Figure 4: Speed-Power curves from the analysis by NAPA

The extensive measurement campaign conducted has had an important role in convincing the market of the functionality of the Rotor Sail solution and its impact. It also provided means to adjust and verify the performance model of the 18x3 Rotor Sails.

The results of the campaign have given confidence to proceed with scaling the Rotor Sail designs to two larger models; the 24 m high and 4 m diameter and 30 m high and 5 m diameter Rotor Sails.

3.2 SIMULATION MODEL

Wind propulsion is relatively new and unknown to the modern shipping industry, and to gain confidence of the market, it is crucial to give realistic indications of the true savings potential. Another typical feature of the wind propulsion is that it's feasibility and economical profitability depends on the area of operations. These two reasons result in need to analyse the feasibility of the Rotor Sails on a case by case basis. It is then crucial to be able to make computationally inexpensive calculations of the performance potential of the Rotor Sails. This enables making of low-threshold case studies for potentially interested customers.

As stated in the previous chapter, the performance model was validated in the trials onboard MV Estraden. With the validated performance model, Norsepower has developed a simulation tool to enable estimating the potential for fuel saving and emissions reduction for specific vessels and routes. The simulation is based on long-term wind statistics, the verified performance characteristics or Rotor Sails, and ship specific operational parameters. The results include estimates for the average produced thrust expressed in main-engine equivalent power (kW), fuel saving and emissions reductions, as well as basic investment profitability indicators. Norsepower has performed numerous case studies with the simulation tool, and routinely provides interested customers with the relevant information to assist them in making investment decisions.

3.3 SCALING OF THE MODEL

As mentioned, to ensure that all real-life physical phenomena are taken into account, measurement data of actual operation is important. This is true not just for the first pilot, but also every time when new size of Rotor Sail is tested as it is not trivial that for example Reynolds number effects are negligible. At the moment, Norsepower has two ongoing measurement campaigns to validate the performance model for 24x4 and 30x5 Rotor Sail models.

3.4 CRUISE FERRY VIKING GRACE

The first Rotor Sail installation on a passenger vessel was completed in April 2018. *M/S Viking Grace* was built in 2013, and it is the first passenger vessel to use LNG fuel. Viking Line has been a forerunner of improving the environmental performance of its vessels, and Viking Line has realised the potential of wind propulsion to further reduce emissions and fuel consumption.

The Rotor Sail installed on Viking Grace is 24 m high and the rotor diameter is 4 m. An extensive measurement campaign for verifying the fuel saving performance is ongoing, and results will be available in 2019. Three independent parties are collecting measurement data for analysis to determine the average fuel saving performance of the Rotor Sail. Another important aspect of the first year of operation is to gather the experiences and comments from the crew, technical department, and passengers. The first eight months of operation have been very encouraging, and the Rotor Sail seems to attract positive attention from passengers.

Figure 5: Viking Line's Viking Grace sails daily between Turku, Finland and Stockholm, Sweden.

3.5 MAERSK P-CLASS TANKER

In partnership with Maersk, Shell, and the UK's Energy Technologies Institute (ETI), the Maersk Pelican, a 109,647 deadweight tonne (DWT) oil product tanker was retrofitted with two 30 m high and 5 m diameter Norsepower Rotor Sails in the end of August in 2018. The project is the first installation of wind-assisted technology on a product tanker vessel. The Rotor Sails are estimated to reduce the average fuel consumption on typical global shipping routes by 7-10%.

An extensive measurement campaign and evaluation of the effectiveness of the Rotor Sails is taking place to test the long-term financial and technical viability of the technology. Independent experts from Lloyd's Register's (LR's) Ship Performance team are acquiring and analysing the performance data during the test phase to ensure an impartial assessment before technical and operational insights as well as performance studies are published.

While the numerical results are still being processed, there are some qualitative early results that can be mentioned.

First is the general notion of the functionality: When wind conditions have been favourable, the Rotor Sails installed onboard Maersk Pelican have been used to propel the vessel forward. This has led to less use of main engine power and/or increased speed of the vessel, contributing to optimising fuel efficiency.

Second concerns the usability of the system. Based on our dialogue with the crew onboard Maersk Pelican, they have indicated they are positive in terms of the use of the technology.

This project is an important breakthrough for the tanker market, which is considered to be a key market where Rotor Sails can provide a competitive solution for achieving the required environmental performance in an economically beneficial way.

Figure 6: Maersk Pelican in port after installation of two 30x5 Norsepower Rotor Sails.

3.6 GLOBAL POTENTIAL

With the help of the Rotor Sail performance model, Norsepower has developed a simulation tool to be able to estimate the potential for fuel saving and emissions reduction for specific vessels and routes. The simulation is based on long-term wind statistics, the verified performance characteristics or Rotor Sails, and ship specific operational parameters. The results include estimates for the average produced thrust expressed in main-engine equivalent power (kW), fuel saving and emissions reductions, as well as basic investment profitability indicators. Norsepower has conducted large number of case studies and the studies correlate well with the busiest global shipping routes.From this database of case studies some remarks can be made. First of all, the area and the route affect both a great deal on the savings potential, which is of course natural as the wind conditions can vary significantly between different areas. Secondly, the most promising areas for Rotor Sails are the routes far from the equator. With respect to global shipping routes this means North Atlantic crossing, North and Baltic Sea areas, and North Pacific crossing.

The best legs in the accumulated data show an annual average savings potential of more than 700 kW with median savings of around 300 kW (speed correction is applied to data to allow comparison of different ship cases, numbers given here are assuming a ship speed of 15kn). This means that the upper half of the simulated routes so far are such that depending on the ship, the annual savings potential with one 30x5 Rotor Sail is 3-5%. If multiple Rotor Sails are installed, the savings potential can be multiplied with the number of units assuming they are well exposed to wind. As a rule of thumb, the distance between Rotor Sails or to other wind shading structures should be at least the height of the Rotor Sail to avoid efficiency reducing turbulence. In general, one can say that there is a significant amount of globally busy shipping routes, that have excellent or very good savings potential.

4. CONCLUSION

New materials and engineering tools have been critically important in developing the Norsepower Rotor Sails to be technically robust, efficient, and economically feasible. Reducing emissions of global shipping operations by 50% by 2050 is an ambitious target. Achieving the target will require application and development of both existing and new innovations and technology, as well as an open mind from all parties. Norsepower offers proven Rotor Sail technology to support the shipping industry in designing and building efficient vessels, which are compliant with existing and forthcoming environmental regulations.

The profitability of installing Rotor Sails depends on the wind conditions on the ship's route and various other operational parameters. There is a significant market with high economical savings potential with today's production prices. In the future it is likely, that the business case for installing Rotor Sails is improved with reducing production costs and increasing fossil fuel related operating costs. This leads to an increasing part of the global shipping routes to become economically feasible for auxiliary wind propulsion.

Auxiliary wind propulsion is one of the key technologies, which already today provides a solution to reduce fuel consumption and emissions. By combining complementing energy efficiency technologies, shipowners are able to build ships which comply with new regulations.

Ville Paakkari holds the current position of Sales Engineer at Norsepower Oy Ltd. He is responsible for performance estimation analysis of the Rotor Sail solution. His previous experience includes Masters' degree in fluid dynamics.

CASE STUDY: WIND-ASSISTED SHIP PROPULSION PERFORMANCE PREDICTION, ROUTING, AND ECONOMIC MODELLING

N J van der Kolk, Delft University of Technology, NL
G Bordogna, Delft University of Technology, NL
J C Mason, The University of Manchester, UK
P Desprairies, Bienfair Inc.
A Vrijdag, Delft University of Technology, NL

SUMMARY

In this paper, a vessel model for the performance of wind-assisted ships is combined with a routing tool to assess the fuel savings available from the installation of both one and two Flettner rotors when travelling along a Great Circle Route path. This is combined with an economic analysis to assess commercial viability for these hybrid concepts. The case study is performed in collaboration with DAMEN shipyards, who have provided a design for a wind-assist concept to sail in the Baltic Sea, that, since January 2015, is an Emission Control Area where a sulphur limit content of 0.1 % is enforced on the ship fuels. Results for this case study are presented in terms of fuel savings and payback period analysis, where the reference case is an identical ship sailing without wind propulsors. For the 5,150 dwt general cargo vessel travelling at a speed of 10 knots, average fuel savings of 2.99% were obtained in the Baltic Sea for the single Flettner scenario, and 6.11% for the double Flettner scenario. A discussion of key engineering and design constraints for these ships is included.

NOMENCLATURE

AWS	Apparent wind speed - ship frame (kts)
CLR	Center of lateral resistance (m)
EEDI	Energy Efficiency Design Index
k	Flettner rotor spin ratio (-)
FC	Fuel consumption (kg/hr)
SFC	Specific fuel consumption (kg/kWh)
TWA	True wind angle - ship frame (deg)
TWD	True wind direction - global frame (deg)
TWS	True wind speed - global frame (kts)
ECA	Emission Control Area
HFO	Heavy Fuel Oil
WASP	Wind-assisted ship propulsion

1. INTRODUCTION

Around 90% of world trade is carried by the global shipping industry, accounting for nearly 3% of anthropogenic carbon dioxide emissions (Smith et al., 2015). (Nelissen et al., 2016) reports that greenhouse gas emissions due to seaborne transport are expected to increase by 50-250% by 2050. Furthermore, Heavy Fuel Oil (HFO), which contains high sulphur content, is still commonly used by the global fleet. This causes the shipping industry to be responsible for around 13% and 15% of the global SOx and NOx emissions due to human activities respectively (Smith et al., 2015). Because of this, the maritime industry is currently under pressure to mitigate its negative impact on the environment and on human health and this is sought by issuing ever stricter emission regulations.

As stated in (Argyros, 2015), wind assistance has the potential to deliver double-digit fuel savings and, according to the same study, it is one of the few technologies for which a substantial cut in fuel consumption is achievable in the short-term. Similar studies performed by (Naaijen et al., 2010), (Traut et al., 2014) Eggers, 2016), (Fujiwara et al 2005), also demonstrate the emission reduction potential of wind assist technology. Given that (Traut et al., 2018) demonstrate that full decarbonisation is necessary by close to the middle of the century for the shipping industry to stay in-line with the Paris agreement, the implementation of short-term emission reduction measures is of vital importance to curb cumulative emissions. Wind propulsion is also one of a limited number of suitable mitigation options that provide a reduction in both carbon dioxide and other pollutants, such as SOx and NOx (Bows-Larkin et al., 2014). Its implementation could help to avoid a potential lock-in of infrastructure that solely focuses on current sulphur regulations, such as with Liquefied Natural Gas.

In this paper, a modelling for the performance of wind-assisted ships is combined with a routing tool for voyage optimisation and an economic analysis to assess commercial viability. The case study is performed on two Baltic Sea routes, a North-South route between Lulea and Gdańsk, and an East-West route between St. Petersburg and Stockholm. The study is conducted with the support of DAMEN shipyards, which has provided vessel details for the Combi Freighter 5000, a coaster. The economic analysis is presented in terms of a payback period analysis.

This paper presents the work of contributors across several disciplines. An outline is provided to orient the reader: The paper begins with a review of operational, regulatory and market drivers for the greening of sea shipping. Next, the details of the present case study are provided, including the statistics for wind conditions for each route and the wind propulsor investment. The methodology section first described the approach whereby the analysis efforts of each contributor are combined, followed by a (necessarily brief) description of the individual models. The results are presented in three parts: First, the vessel modelling is reviewed with attention placed on key engineering and design constraints for wind-assist ships, followed by a discussion of the weather routing results, and the payback period analysis for both cases. Finally, a description of the ongoing work of each contributor is included in the conclusion.

2. STRONG DRIVERS TAKE SEA SUPPLY CHAINS TOWARD GREENER FUTURES

While volatile fuel costs and constraining environmental regulations push sea freight operators toward low carbon operational business models, growing retail consumer awareness toward current environmental challenges also push carrier operators toward a cleaner supply chain, whereby a greener shipping fleet can deliver a higher value to their client. In this economical discussion, we enumerate three significant drivers that will prompt sea freight operators to shift toward a greener shipping industry.

2.1 OPERATIONAL DRIVERS

Fuel costs represented 40% of the sea freight industries overall operational costs in 2010 (Samitas, 2010). This figure has increased up to 50-60% in 2018 (Stratiotis, 2018) due to the recent steep increase in fuel prices and also because of the increasing use of alternative fuels. As an example, the upcoming introduction of the 2020 Sulfur Cap, with its expansion of Emission Control Areas to a global scope, will result in the increased investment in expensive alternative fuels such as LNG by sea freight operators, with an estimated impact of $50,000 a day for a 20,000 TEU ship (Stratiotis, 2018).

Increasing fuel prices is also accompanied by an increase in market volatility. 2018 saw the CBOE Crude Oil Volatility Index peaking to a 10-years high, resulting in fuel prices that had not been as volatile since the 2008 financial crisis. As a consequence, the sea freight industry has been using financial instruments, such as swap or futures in order to hedge from volatile bulk prices, implying a premium equally as expensive than the chosen hedging level (Samitas, 2010).

2.2 REGULATORY DRIVERS

The International Maritime Organisation (IMO) acknowledges that energy efficiency measures are largely underutilised and that market drivers alone are insufficient to bring about the required technical and operational measures (2nd IMO report 2009, Acciaro et al, 2011). The IMO estimates that unrealised potential for efficiency savings in the maritime sector lies between 25% and 75% for CO_2 emissions. The shipping industry, relatively unregulated when compared to other elements of the global economy, is now subject to two IMO policies for ships: the Energy Efficiency Design Index (EEDI), imposing increasingly ambitious targets for new-build ships, and the Ship Energy Efficiency Management Plan (SEEMP), intended to improve the operational efficiency of all ships. The EEDI targets are to increase from 10% (present) to 30% in 2025.

It has been shown that projected growth for the maritime shipping sector is such that even a full implementation of presently available technology, under existing policy measures, would be insufficient to curb the contribution of the sector to the sum of anthropogenic CO_2 emissions (Bazari, 2011). It is for example unlikely that the targets of the European Commission (40-50% reduction by 2050) will be reached, as demonstrated by (Faber et al., 2009) and (Acciaro et al, 2011). It has also been reported that the EEDI is not adequately incentivising the adoption of all innovative measures with the potential for significant savings (IMO 2015)

2.3 MARKET DRIVERS

Wind-assist commercial shipping also represents a market opportunity, potentially increasing the perceived value of the shipped products. Indeed, a Nielsen survey reported a growing demand for greater sustainability from end-consumers, with 66% of global consumers willing to pay more for sustainable brands in 2015 (up 55% from 2014), although this result needs to be tempered considering the fact that only small-sized niche markets actually do buy sustainable products (Littler, 2011). Since WASP technology makes supply chains greener, it can consequently be highly valued to supply these niche markets.

Beside, wind-assist commercial shipping can also be used toward stronger and wider marketing strategies. Green chains favour chain transparency and vice-versa (Mol, 2015), with transparency increasing the brand trust and eventually strengthening some aspects of the consumer-brand relationship such as loyalty (Kang, 2013). But there is even more potential to that. Indeed, transparency enables Business-to-Consumer companies to collect, index and retrieve supply chain information in the form of stories (Woodside et al., 2008), implementing storytelling as an efficient marketing strategy (Pulizzi, 2012). Quoting Michael Margolis in 2014, "people don't buy a product, service, or idea; they buy the story that's attached to it." Wind-assist commercial shipping can provide additional marketing substance to sea freight operators' clients, bringing novel value, and the possibility to sell the shipped products at a higher price to the final consumer.

In Summary:

- Increasing fuel costs is expected to incite sea freight operators to implement low-consumption changes.
- Based on the growing understanding that present regulatory measures are insufficient, a more stringent regulatory climate is expected.
- Sustainable supply chain brings additional value to final consumers and raises the possibility of increasing the perceived value of the shipped good.
- To be watched: OPEC strategy to lower fuel price, deregulation moves.

3. CASE STUDY

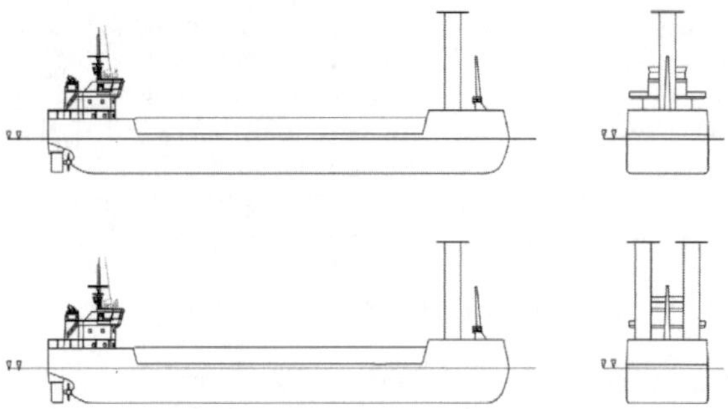

Figure 1: CF5000 with one (top) & two (bottom) Flettner rotors

3.1 THE SHIP

The ship considered in the present case study is a general cargo coaster, the Combi Freighter 5000, whose specifications were given by DAMEN shipyards. The ship has the following characteristics: LOA= 86.82 [m], Beam= 15.20 [m], Draft= 6.35 [m], Displacement= 6682 [t] and Deadweight= 5150 [t].

3.2 WIND PROPULSION SYSTEMS

Two different solutions were considered, a single Flettner case with one Flettner rotor and a double Flettner case with two Flettner rotors (see Figure 1).

Specification	Value	Units
Diameter	3	[m]
Span	18	[m]
Weight (including foundation)	29	[ton]
Buying cost	300k	[euros]
Installation cost	50k	[euros]
Maintenance cost	7k	[euros/year]
Lifetime	25+	[years]

Table 1: Details of the Flettner rotors used in the case study

The wind-propulsion system used to retrofit the analysed vessel is the Flettner rotor. In particular, a small size Flettner rotor, as offered by the company NORSEPOWER, was chosen. The details of the selected system are given in Table 1.

3.3 BALTIC SEA ROUTE

Figure 2: Map of the Baltic Sea. (Google Maps, 2019)

The ship will be operating in the Baltic Sea region. The route has been simplified to two legs, an East-West route between St. Petersburg and Stockholm, and a North-South route between Lulea and Gdansk (see Figure 2).

Route	Point 1	Point 2	Distance
Lulea-Gdansk	21.75E, 65.25N	18.75E, 54.75N	1179km
St Petersburg-Stockholm	29.25E, 60N	18.75E, 59.25N	596km

Table 2: Details of the routes selected for analysis in the Baltic Sea region.

The details of the routes, including the latitude and longitude point of each destination, and the total distance between the routes can be seen in Table 2

Information about the True Wind Speed (TWS) and True Wind Direction (TWD) was collected along both routes using ERA-Interim weather data obtained from the European Centre for Medium-Range Weather Forecasts (ECMWF), and this has been presented as a wind rose plot for each leg (Figures 3 and 4). Wind rose plots are polar plots displaying both wind speed and direction. The radial direction represents the TWD and the colours represent the TWS. A TWD of 90 degrees represents a wind blowing from west to east, or a westerly wind. The wind speed is binned according to wind strength. The thickness of each colour indicates the frequency that the wind speed occurs.

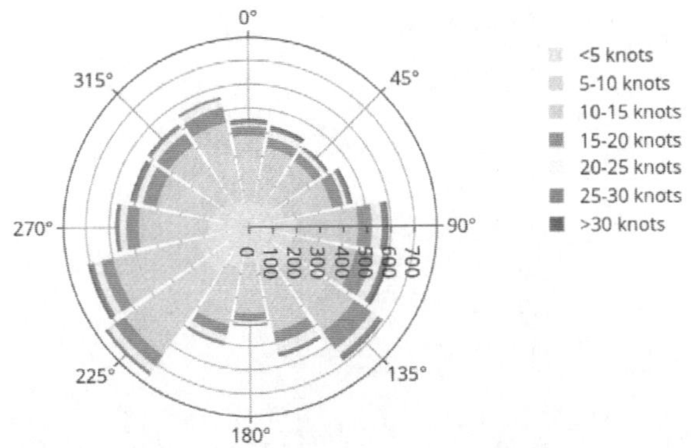

Figure 3: Wind rose plot for the route between Lulea and Gdansk.

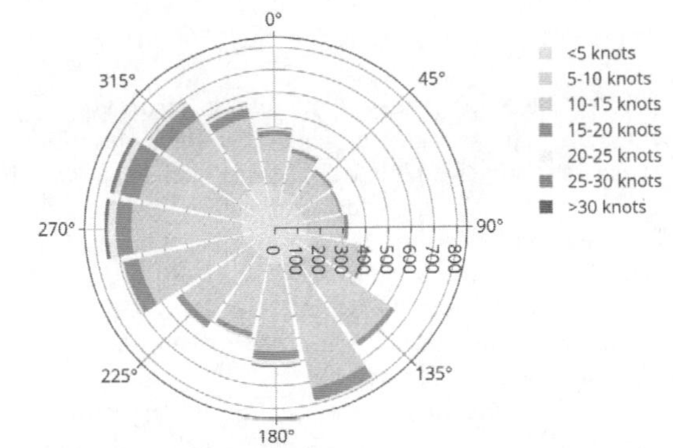

Figure 4: Wind rose plot for the route between St. Petersburg and Stockholm.

From the wind rose plots, it can be seen that the wind conditions for both routes are light and variable. The TWS experienced on the North-South route between Lulea and Gdansk is mainly between 0 and 15 knots. The dominant wind direction on the route is 225 degrees (a north-easterly wind), with a secondary peak at 135 degrees (a north-westerly wind). Along the East-West route between St. Petersburg and Stockholm, the dominant wind direction is easterly, with a secondary peak for north-westerly wind. The maximum TWS for both routes is over 30 knots, but this occurs only a small proportion of the time. In particular, for the East-West route between St. Petersburg and Stockholm, this occurs less than 0.3% of the time.

4. METHODOLOGY

4.1 VESSEL MODEL

To arrive at the achievable fuel savings of the considered wind-assisted ship sailing along a desired route, it is first necessary to study its performance in terms of aerodynamics, hydrodynamics and total fuel efficiency. Key components of the vessel model are indicated in Figure 5.

The use of wind propulsion systems to transform wind energy into forward thrust may be used in two ways:

- Increase the speed of the ship while maintaining the same engine output the ship would have without wind-assisted propulsion
- Reduce the use of the main engine while maintaining the same service speed the ship would have without wind-assisted propulsion

The latter solution is generally considered the most interesting from an operational perspective, and is adopted for the vessel modelling presented here.

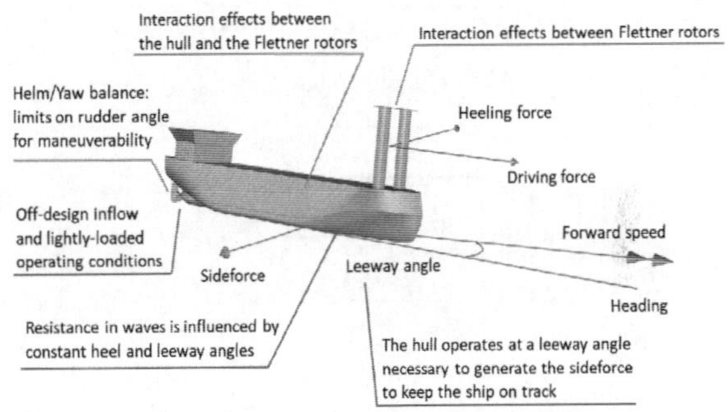

Figure 5: Key components of the vessel model.

The desired wind-generated thrust also introduces undesired aerodynamic forces that need to be balanced by corresponding hydrodynamic reactions. For a chosen vessel service speed, the equilibrium between the aerodynamic and hydrodynamic forces and moments is found with the use of an optimisation routine that selects an equilibrium with the minimum required propeller thrust. The system is solved in four degrees of freedom: surge, sway, roll and yaw. The propeller thrust that the ship must generate to achieve an equilibrium is calculated as the sum of all components in surge:

T(1-t) = Wind assist thrust -
Calm water resistance -
Added resistance in waves -
Rudder induced resistance -
Hull induced resistance -
Hull windage

The results of this procedure are then passed to the fuel consumption model. In fact, as hybrid vessels, the performance of a wind-assist design will depend on the contribution of the wind-propulsion system alongside the efficiency of the main engine, where (in general) an off-design operating point will be detrimental for the specific fuel consumption. Finally, the power required to rotate the Flettner rotors is taken into account as an added load for the ship electric grid.

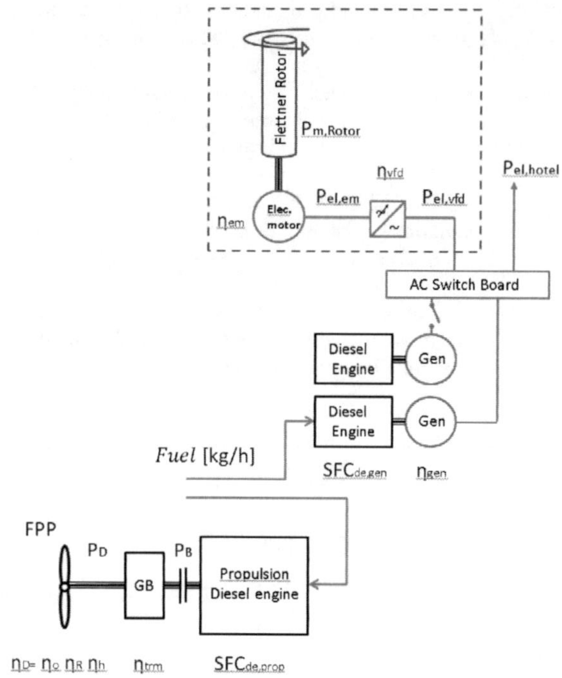

Figure 6: Schematic of the main propulsor and ship electric grid.

The power and propulsion system that is considered in this paper is visualised in Figure 6. It shows a fixed pitch propeller, driven by a main propulsion diesel engine, which delivers brake power P_B. The fuel in [kg/h] required to deliver this brake power depends on the operating point of the engine in terms of P_B and $SFC_{de,prop}$. The top right part of Figure 6, shows the electrical (AC) grid of the vessel. It shows two installed gensets, of which one is online. The online genset feeds the switchboard, from which the electrical consumers are fed. In this case we focus on one particular consumer: the Flettner rotor, which is driven by an electrical motor, which in turn in fed via a variable frequency drive to allow for control of the Flettner rotor rotation speed. Since this drive system requires power it should be taken into account when estimating the potential of the total WASP system to reduce overall fuel consumption.

The solution of the process here described, that is the fuel consumption in [kg/hr], needs to be determined for all combinations of true wind angles and true wind speeds that the ship might encounter en route. This is the backbone idea behind the vessel modelling.

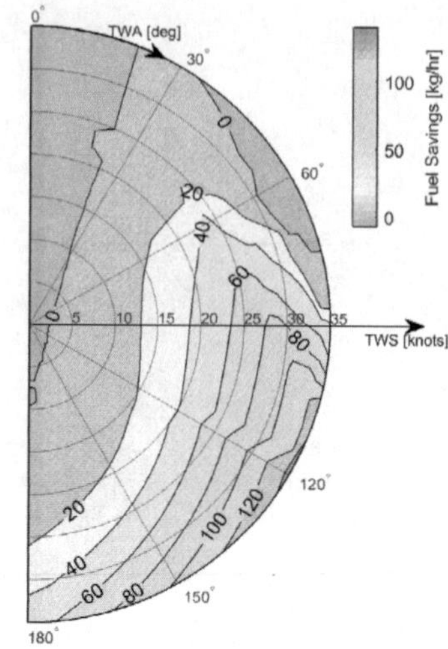

Figure 7: Fuel savings polar for single Flettner case

Results for vessel modelling are typically presented in a polar plot, in which variables describing vessel performance are plotted for wind speeds and wind directions. The axes are defined relative to the ship heading: the ship is sailing directly into the wind at TWA = 0 deg, and with wind astern for TWA=180 deg. In the polar diagram presented in Figure 7, the fuel savings were computed by taking the difference between the Single Flettner case and the reference case.

In the remainder of this section, a description of the modelling for each aspect of the vessel modelling is provided. Given the scope of work presented, the level of detail is necessarily brief. References are provided to relevant technical publications.

4.1 a) Aerodynamics

From an aerodynamic perspective, to achieve a reliable performance prediction, it is necessary to consider the selected wind-propulsion system in its real sailing conditions. This means that scale effects, interaction effects occurring between the several wind propulsors installed on the ship's deck, as well as the interaction effects occurring between the propulsors and the ship itself, need to be taken into account. The aerodynamic interaction effects entail a change in wind speed and wind angle of incidence that will generate a change in the lift and drag forces generated by all the wind propulsors employed and, for these reasons, cannot be neglected.

In this work, the Flettner rotor is the wind-propulsion system chosen to be installed on the analysed vessel. The Flettner rotor is a cylinder spun by an electric engine and it generates a lift force due to the Magnus effect. Nowadays, Flettner rotors are becoming increasingly popular as, in change of a minimal absorbed power, they generate a significant amount of thrust with a reduced "sail" area.

The aerodynamic data used in the current study are the results of dedicated experiments carried out at Politecnico di Milano wind tunnel. In particular, two separate series of tests on Flettner rotors were conducted. The first test series dealt with a large-scale rotating cylinder, in which the objective was to measure the aerodynamic forces for different Reynolds numbers in order to study possible scale effects (Bordogna et al., 2018). The highest Reynolds number reached was $Re=10^6$, that is roughly between one and four times lower the actual Reynolds numbers experienced by a full-scale Flettner rotor in typical sailing conditions. The CL and CD data relative to the results obtained at $Re=10^6$ were used throughout this work as baseline.

The second series of tests dealt with two smaller Flettner rotors, set at several different combinations of distances and angles with respect to the incoming wind. These experiments aimed at measuring the effects of the aerodynamic interaction on the lift and drag forces generated by each Flettner rotor. The results of such experiments were used in the current work to modify the lift and drag obtained in the first test series, given the relative distance, position and spinning velocity of each installed Flettner rotor.

Lastly, the effects of the interaction between the ship's hull and the Flettner rotors, and the windage of the ship were calculated according to (Walree, 1988) and (Fujiwara et al., 2005) respectively.

4.1 b) Hydromechanics

Fitting a commercial vessel with an auxiliary wind propulsor will introduce a set of forces and moments besides the desired aerodynamic thrust. The ship will adopt a steady heel and leeway angle -- the sailing condition -- to arrive at an equilibrium for each operating condition. The leeway angle is the angle-of-attack for the hull, necessary to generate a hydrodynamic sideforce in opposition to transverse component of the aerodynamic force. Further, the distribution of the hydrodynamic sideforce along the hull will result in a net yawing moment. At last, the vertical separation between the sideforce components will create a heeling moment.

The performance of a wind-assist concept will depend on the contribution of the wind propulsor to the required thrust, alongside the efficiency of the sailing ship. The underwater ship of a conventional freighter is ill-suited for efficient sideforce production, and significant induced resistance is to be expected. Of course, the introduction of a sail-plan will only benefit the vessel if the net thrust gained outweighs any loss in efficiency or increase in resistance. The course-keeping ability of a ship represents a key modeling and design challenge for wind-assist vessels. A conventional cargo vessel hull—with an undersized rudder as the only appendage—is essentially unfit for sailing. Such a hull will operate with greater leeway angles, and with 'weather helm' as a consequence of this inefficient side-force generation. The rudder can be used to oppose this destabilising moment, with an associated resistance penalty. Most of all, the vessel must be able to maneuver. This is a fundamental operational constraint that must be present in any vessel modelling. The sailing efficiency and course-keeping are determined with a Reynolds-Averaged Navier Stokes computational fluid dynamics (RANS-CFD) tool. The simulation methodology for analysis of wind assisted ships has been validated in (van der Kolk et. al. 2019) and (van der Kolk et. al. 2019).

Hydromechanics for the reference cases (ship operation without wind assistance) follows standard industry practice. The ship resistance curve is determined with the RANS-CFD simulation tool. The contribution of the added resistance in waves is determined with the method of Gerristma-Beukelman (Gerritsma & Beukelman 1972) using sectional hydrodynamic loading. Under the present analysis, it is assumed that the sea state is coupled to the wind direction and strength. The sea state is defined with a Pierson-Moskowitz spectrum for fully developed seas. The spectrum is scaled according to the Beaufort scale for significant wave height, where a maximum value for the wind speed is prescribed (20 m/s) to avoid unrealistic fully developed storm sea.

4.1 c) Fuel consumption model

To make a fair comparison of total fuel consumption between the baseline ship and the WASP designs (schematically visualised in Figure 6), the following approach was taken: First of all, a realistic baseline propulsion system design was made, resulting in a maximum ship speed of 11 knots at maximum engine speed and 85% MCR For both ships a fixed pitch propeller was assumed. Subsequently, for both ships, for a fixed ship speed of 10 kts (adopted for the remainder of the analysis), the required propeller thrust for each combination of true wind speed and direction was determined, taking into account the contribution from resistance penalties incurred as a result of the sailing operating condition of the ship.

For each combination of wind speed and angle, the required propeller speed and power is determined, using a Wageningen B propeller open water diagram, assuming that the increased drift angle of the WASP ship does not influence the propeller behaviour significantly via the disturbed wakefield. Based on these results, in combination with a gearbox reduction ratio and transmission losses, the engine operating point per condition is determined, including a "flag" in case the required engine operating point is impossible because it lies outside the engine operating envelope, which can for instance happens in case of strong headwinds. A parametric model of specific fuel consumption (SFC in kg/kWh) of the propulsion diesel engine (Shi, 2010) is incorporated from which, for each operating condition (given required engine speed and brake engine power), a fuel consumption in kg/h, required for propulsion can be calculated.

For the WASP case, the power required to spin the Flettner rotor, has to be taken into account, since this results in additional electrical load, and therefore into additional fuel consumption. The efficiency train from required mechanical Flettner rotor power to generator fuel consumption contains an electric motor, a variable frequency drive, and a generator set, as shown in Figure 6.

Note that the SFC behaviour of the genset is not modelled in detail as is done for the propulsion diesel engine: instead a constant SFC is taken which is typically higher than the SFC of the larger, slower running propulsion diesel engine.

4.2 WEATHER ROUTING

Using the fuel consumption model discussed above, it is possible to obtain the fuel consumption of the vessel, in kg of fuel per hour, for a given wind direction and speed. These results are then passed to the routing algorithm.

The weather routing module is broken down into two components : a Great Circle Route (GCR) simulation and a voyage optimisation simulation. This allows a relatively fast GCR simulation to be performed independently of the computationally intensive voyage optimisation simulation.

The GCR is composed using equations developed in a planar coordinate system by (Chen et al., 2015). This allows the latitude and longitude of multiple waypoints on the GCR to be calculated. These waypoints were obtained for the outward and return routes between Lulea-Gdansk and St Petersburg-Stockholm. For each route, the GCR was split into six waypoints, corresponding to five stages. At each waypoint, the wind speed and direction is found using ERA-Interim data from ECMWF with 0.75X0.75 degrees spatial resolution. From this, the angle between the wind and the vessel heading is found. These values are used in conjunction with the fuel consumption modelling, to calculate the fuel consumed between adjacent waypoints. Subsequently, the fuel consumed on the entire route can be found.

The GCR routing methodology presented here assumes that the vessel perfectly follows the GCR, requiring a constant bearing change. In reality, this may not be possible, and the vessel may approximate the GCR by following several rhumb lines (lines of constant bearing). This would produce a discrepancy in the wind direction experienced by the vessel. However, due to the small distances considered in this study, these discrepancies would be minimal and the GCR routing will provide a good approximation.

Voyage optimisation incorporates the weather conditions experienced by the vessel on the route to reduce the total fuel consumed on the voyage. This involves altering the route of the vessel, away from the GCR, to search for advantageous weather conditions. Dijkstra's optimisation algorithm (Dijkstra, 1959) is commonly implemented to minimise the fuel consumption along the route, and the algorithm is a grid-based algorithm. An associated cost of travelling between adjacent grid waypoints is set according to the fuel consumed when travelling between the two points. Based on these associated costs, the optimisation algorithm searches all possible pathways along the grid and determines a route with minimal fuel consumption.

This study solely focuses on simulations of the wind-assisted vessel on the GCR, with voyage optimisation subject to future research. The outward and return journeys of both routes were simulated for the 5,150 DWT general cargo vessel travelling at a speed of 10 knots. The routes were simulated for the vessel with no wind propulsion installed, for the vessel with one Flettner rotor installed and for the vessel with two Flettner rotors installed. Subsequently, the fuel consumption savings from the addition of both one and two rotors were found relative to the vessel with no wind propulsion.

4.3 ECONOMIC ANALYSIS

In order to assess the economic viability of the wind-assist technology in a commercial perspective, the payback period of the project (whose duration is assumed to be the average lifetime of maritime equipment) needs to be assessed assuming a 300 months-period. To do so, the costs over time induced by the WASP project are compared with the costs over time for the reference case.

The payback period is defined through two main cost functions:

- Investment[€] : overall investments needed to buy and install WASP equipment.

- Savings(t)[€/t] : The changes in overall cash flow savings resulting from the installation of the WASP technology, calculated on a monthly basis.

The Savings function is defined as the difference between the total operational costs under a WASP installation scenario and the total operational costs without WASP installation scenario, all other things being equal. After breaking down the operational costs into several sub-cost functions, it is also possible to write down the Savings function as the sum of all these sub-cost functions between the two scenarios :

$$\begin{aligned} \text{Savings(t)[€/t]} = \ & \Delta\text{fuel cost(t)} + \\ & \Delta\text{maintenance(t)} + \\ & \Delta\text{revenue(t)} + \\ & \Delta\text{insurance(t)} + \\ & \Delta\text{administration(t)} + \\ & \Delta\text{payroll(t)} \end{aligned}$$

4.3 a) Fuel Savings: Δfuel cost(t)

The change in operating costs due to fuel consumption (WASP technology vs Business as Usual) is the result of the vessel and routing modelling detailed above. The result is a statistical average for fuel savings per month. An operating profile (days at sea per month) of 0.72 is assumed. Finally, p[€/ton] is defined as the price for fuel. For operation in the Baltic Sea region, a designated Emission Control Area, the ship is obliged to purchase 0.1% sulphur grade fuel. The fuel price is a key driver for the economic model and the sensitivity to the value is studied. The fuel cost function now reads:

$$\Delta\text{fuel cost(p,t)} = \text{p[€/ton]} * \text{fuel savings [ton]} * \text{Operating profile [-]}$$

4.3 b) Added maintenance costs: Δ_maintenance(t)

According to NORSEPOWER, the installation of wind-assist technology implies maintenance costs amounting to 2% of the initial investment costs annually. Therefore, the maintenance cost function is as follows:

$$\Delta\text{maintenance}(t) = -2\% * \text{Investment} * t/12$$

4.3 c) Revenue variation: Δrevenue(t)

The added weight of the Flettner rotor installation will reduce cargo deadweight by up to 0.9%. Insufficient information was available to compute the impact of this variation in the cost, but in light of the magnitude, it is assumed that this variation is negligible. Similarly, we will not mention the handling costs variation in harbours coming from the deadweight variation.

4.3 d) Insurance and administration costs variation

It is assumed that insurances and administration costs do not vary.

4.3 e) Payroll costs variation

According to NORSEPOWER, there is no need to hire additional crew, meaning that there is no payroll cost variation.

Finally, the sum of all variations considered is written as:

$$\text{Savings}(p,t)[€/t] = \Delta\text{fuel cost}(p,t) + \Delta\text{maintenance}(t)$$

The payback period is found by determining the number of months, t, for which the savings function equals the initial investment:

$$0 = -\text{Investment}[€] + \text{Savings}(t,p)[€/t]$$

5. RESULTS & DISCUSSION

5.1 VESSEL MODEL

Before considering the results in terms of fuels savings, it is of interest analysing the polar diagrams of some parameters that are important for the sailing behaviour of the vessel. Significant effects are highlighted for each component of the vessel model.

5.1.a) Aerodynamics

From an aerodynamic perspective, it is interesting to compare the spin ratios of the Flettner rotors (that is the ratio between the tangential velocity of the rotor and the incoming wind speed) to understand the effects of the Flettner rotor aerodynamic interaction on the results.

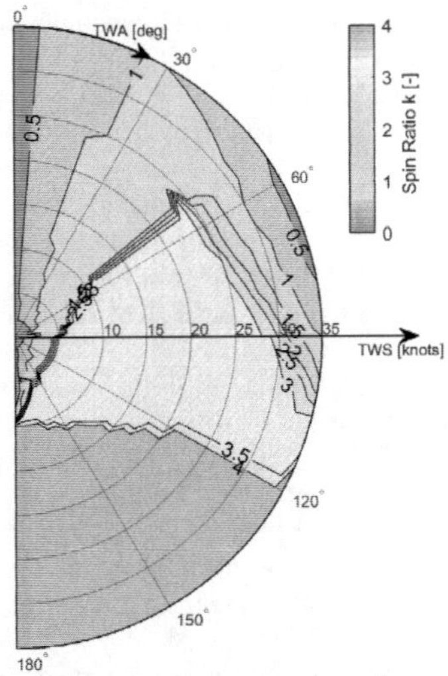

Figure 8: One-Flettner rotor case: spin ratio

The spin ratio, in fact, has a direct influence on the amount of aerodynamic force generated. In Figure 8 the spin ratio for the one-rotor case is depicted, whereas the spin ratio for the two-rotor case is given in Figure 9. In the latter case, the polars are reported both for the port side and the starboard side rotor. Looking at the spin ratio polars, it can be noticed that the two configurations show some similarities but also several differences.

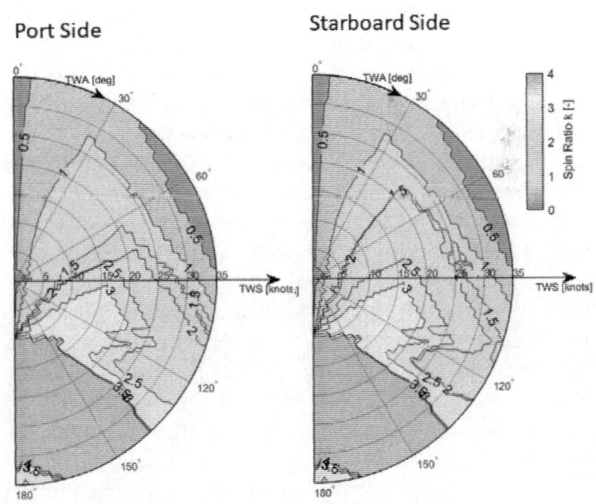

Figure 9: Two-Flettner rotor case: spin ratios

Both designs adopt a low spin ratio when sailing upwind and a larger spin ratio when the wind comes from astern. This is because when sailing at small TWAs is more beneficial to have a high lift/drag ratio, and this occurs at low spin ratios ($k \approx 1$). Conversely, at large TWAs, the forward thrust is generated both by the lift and the drag forces. High values of lift and drag occur at large spin ratios. It should be noticed that a plateau for both the lift and the drag force is found at $k=4$, meaning that a further increase of the spin ratio would only produce a detrimental rise of the power consumption of the Flettner rotors.

Regarding the differences encountered in the two designs, for wind angles 60°<TWA<150°, there is a marked difference between the spin ratios of the one- rotor and the two-rotor design. In this range of wind conditions, in fact, the two Flettner rotors strongly interact with each other, generating (generally) detrimental variations of the incoming flow angles. To compensate for these effects, and to generate an optimum overall thrust, the spin ratios of the port side and starboard side Flettner rotor differ from each other and are (generally) lower with respect to the single-rotor case. The trade-off is that a lower spin ratio means a lower thrust but also less detrimental interaction effects. It should be kept in mind that, in case no aerodynamic interaction effects were modelled, the polars of the port side and starboard side rotor would be the same as the polar of the single-rotor case, meaning that the aerodynamic thrust would have been overestimated.

On the other hand, for wind angles 30°<TWA<60°, the differences in spin ratios between the single and the double rotor design, is caused by an imposed limit on the rudder angle in order to maintain sufficient maneuverability. In this condition, both Flettner rotors of the two-rotor design have to depower (adopt a lower spin ratio) to keep the balance of the ship within the imposed limit for the rudder angle.

5.1 b) Hydromechanics

A survey of the hydromechanic response of the system reveals one key effect : the yaw-balance for the ship. The flow mechanisms responsible for sideforce generation have a destabilising effect on the course keeping ability of the ship. This effect severely limits the operating points of the wind-assist vessel. It is a key design concern.

The physical mechanism responsible for the strong destabilising yawing moment is referred to as the Munk moment. The Munk moment is (in principle) a pure couple that arises as a body encounters an oblique flow. Whereas the yawing moment will develop rapidly, the sideforce production for the hull of a commercial ship is heavily dependant on the higher-order component that is characteristic for low-aspect-ratio lift. The hydromechanic vessel response in the yaw equation is defined by the center of lateral resistance (CLR), a quantity that expresses the distribution of the hydrodynamic sideforce, including the Munk moment. For sailing yachts, these phenomena are experienced as weather helm, a steady rudder angle required to keep a course.

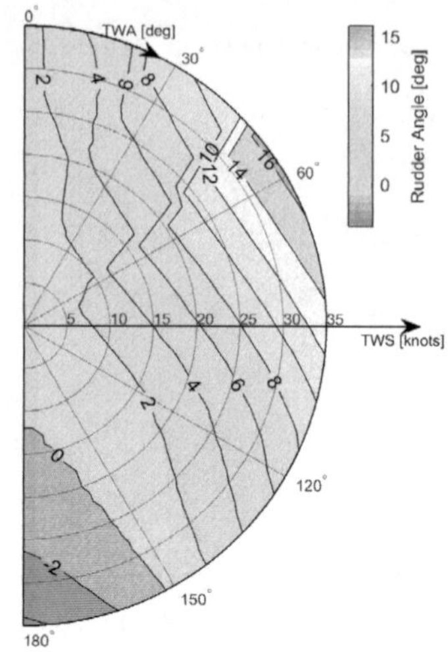

Figure 10: Rudder angle for single-rotor case (no limit imposed)

Commercial ship types, operating without a keel or other specialised appendages designed to mitigate this effect, are essentially unfit to sail. The rudder angle for the single-rotor case is presented in Figure 10 to underscore this point. For upwind sailing conditions at 60° TWA, the necessary rudder angle increases to as much as 16°. The maneuverability of the vessel is clearly impacted for this relatively modest implementation of WASP technology. A limit for the rudder angle has been set to 10°, after which the Flettner rotors will need to depower. This effect is visible in Figures 8 for the single rotor case, where the contour corresponding to 10° rudder coincides with a reduction in spin ratio. Similarly, the rotors in the Two-Flettner case are never fully engaged for upwind courses, as seen in Figure 9. The effect of this depowering is seen in the required propeller thrust as well (Figure 11). For reference, the heel is less than four degrees for the single Flettner and double Flettner cases.

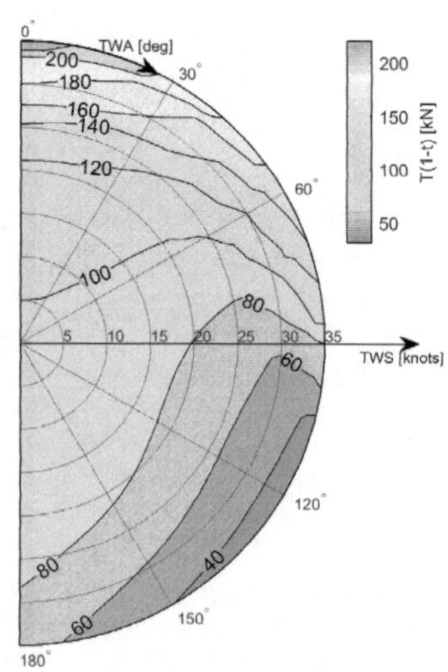

Figure 11: Required propeller thrust

5.1 c) Fuel Consumption Model

Based on the aerodynamic and hydrodynamic force calculation, for each true wind speed and angle, the required propeller force to sail 10 kts can be calculated, both for the reference case and for the WASP case(s). Figure 11 shows this force for the single rotor case. Note that the chosen spin ratio of the Flettner rotor and the restriction on rudder angles has an effect.

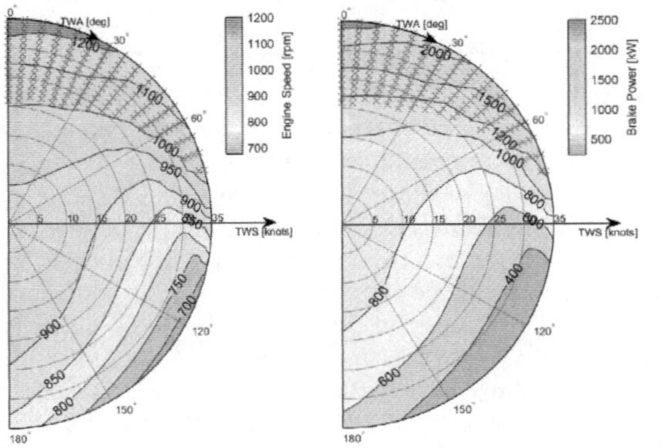

Figure 12 Main engine speed (left) and brake power (right)

Subsequently, the propeller speed, engine speed (Figure 12(left)) and brake engine power (Figure 12(right)) that are required to deliver this force can be calculated, based on the open water diagram and the fixed gearbox reduction ratio. However, the propulsion diesel engine cannot deliver all combinations of engine speed and power: only combinations that lie inside the engine envelope are possible. Combinations that are not allowed are visualised in Figure 12 by the red x-markers.

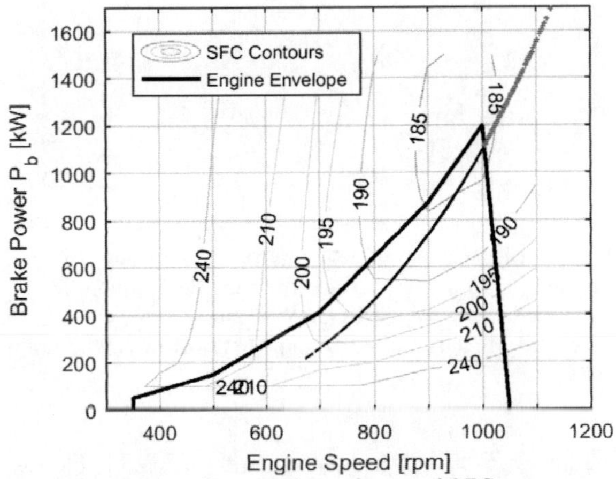

Figure 13 Main engine speed envelope and SFC curves

Their locations can be cross-verified with the engine envelope as shown in Figure 13. These operating conditions correspond to a ship speed reduction due to strong headwinds/seas, which is currently not captured in the routing model. However, for these cases, the reference ship and the wind-assist concepts will be similarly hindered (i.e. the fuel savings will be zero).

Subsequently, for all wind speed and angle combinations, the SFC of the main engine can be determined. To make a fair comparison, additional fuel consumption of the generator due to the power required to spin the Flettner rotor has been calculated and added to the fuel consumption of the main engine. The result of this vessel modelling is expressed as fuel savings compared to the reference case, and has been presented at the beginning of Section 4 (Figure 7). This polar diagram is passed to the routing model.

5.2 WEATHER ROUTING

The Baltic Sea routes were simulated for four departure times per day for the whole year of 2015. The four departure times selected were 00:00:00, 06:00:00, 12:00:00 and 18:00:00, using the weather data obtained from ECMWF. The simulations were performed for the vessel without a Flettner rotor installed, with one Flettner rotor installed and with two Flettner rotors installed.

5.2 a) Lulea - Gdansk

	Lulea - Gdansk		Gdansk - Lulea	
	Fuel savings (%)	Fuel savings (kg)	Fuel savings (%)	Fuel savings (kg)
One rotor	3.67%	373	3.46%	353
Two rotors	7.49%	755	6.96%	703

Table 3: Average fuel savings provided by the installation of a single Flettner and double Flettners per trip, for the outward and return route between Lulea and Gdansk.

The average yearly total fuel savings, in both percentage and kilograms, was calculated for both the Single Flettner and the double Flettner scenarios, using data from the 1,460 simulated voyages from each route. This was done for the outward journey between Lulea and Gdansk, as well as for the return journey from Gdansk to Lulea. The results can be seen in Table 3. Simulations of the Single Flettner scenario produced average fuel savings of 3.67% and 3.46%, whereas simulations of the double Flettners scenario produced average fuel savings of 7.49% and 6.96%. For both journeys, the average fuel savings produced by double Flettners were around double the fuel savings produced by one. Both the outward and the return journey also produced very similar fuel savings when averaged throughout the full year for both the Single Flettner and the double Flettner case.

5.2 b) St Petersburg - Stockholm

	St Petersburg - Stockholm		Stockholm - St Petersburg	
	Fuel savings (%)	Fuel savings (kg)	Fuel savings (%)	Fuel savings (kg)
One rotor	2.78%	141	2.05%	105
Two rotors	6.20%	313	3.79%	194

Table 4: Average fuel savings provided by the installation of a single Flettner and double Flettners per trip for the outward and return route between St Petersburg and Stockholm.

The fuel savings calculations were repeated for the voyage simulations between St Petersburg and Stockholm. The results can be seen in Table 4. Simulations of the Single Flettner scenario produced average fuel savings of 2.78% and 2.05%, whereas the double Flettner scenario produced average fuel savings of 6.20% and 3.79%. Again, the simulation of an additional rotor produced average fuel savings of around double. The journey between St Petersburg and Stockholm also showed higher average fuel savings than the return journey, for both the Single Flettner and the double Flettner case.

The results show that the average fuel savings for the route between Lulea and Gdansk (North-South) are larger than for the route between St Petersburg and Stockholm (East-West), pointing towards the existence of more advantageous weather conditions for wind-assisted vessels along this route. Referring back to the wind rose plots (Figure 3 and Figure 4) in Section 3.3, this result is expected, due to the higher frequency on strong winds on the route between Lulea and Gdansk. The route between Lulea and Gdansk also experiences greater savings than the return route between Gdansk and Lulea. This is again expected, due to the slight presence of a dominant wind from the north easterly and north westerly direction, which is a beneficial wind angle to produce fuel savings from the rotors on the Lulea to Gdansk journey. Similarly, the route between St Petersburg and Stockholm experiences greater savings than the return route between Stockholm and St Petersburg. This is once again expected due to the presence of a dominant easterly wind direction, providing tailwind for the first route and headwind for the second.

These results demonstrate that variations in fuel savings occur between different routes in the Baltic Sea region. However, the N-S and the E-W routes are two contrasting routes in this area, and voyages between ports along similar directions (such as Lulea-Gdansk and Helsinki-Gdansk, which both traverse a N-S direction), may produce similar results. The results also demonstrate that fuel savings correspond well with the wind rose plots for each specific route.

Values for average fuel consumption savings agree with the range of values produced by previous studies on Flettner rotors. However, this range is large, due to the variety of factors that have been shown to affect fuel savings produced by wind-assistance, such as vessel size, vessel speed, route, time period analysed, as well as others. Therefore, care must be taken when making comparisons between specific studies. The closest comparison is discussed by Traut et al. (2014), who analyse the average fuel savings for a 5,500 DWT general cargo vessel, travelling at 8.8 knots from Varberg to Gillingham for the year 2011. (Traut et al., 2014) demonstrate 21% savings for a single Flettner rotor. This value is significantly greater than the fuel savings produced in this study, potentially demonstrating that beneficial weather conditions are experienced on the route between Varberg and Gillingham or showing the significance of the larger Flettner rotor dimensions incorporated in the study by (Traut et al., 2014).

5.2 c) Single Flettner vs. double Flettners

	All Baltic voyages		
	Average fuel saving	90th percentile fuel saving	Maximum fuel saving
One rotor	2.99%	8.22%	34.2%
Two rotors	6.11%	17.8%	89.6%

Table 5: Average fuel savings, 90th percentile fuels savings and maximum fuel savings for both one and two rotors for all combined routes on the Baltic Sea.

The fuel savings from all four journeys were combined to assess the performance of the technology throughout the Baltic Sea for both the Single Flettner and the double Flettner scenarios. From this, the average fuel saving, 90th percentile fuel saving and maximum fuel saving was calculated for the Baltic Sea. These results can be seen in Table 5 and the distribution of values can be seen in Figure 14 for the double Flettner case. The Single Flettner case shows a similar distribution of results. The average fuel saving for the Single Flettner case and the double Flettner case was 2.99% and 6.11% respectively, with the double Flettner again roughly providing double the fuel savings throughout a full year.

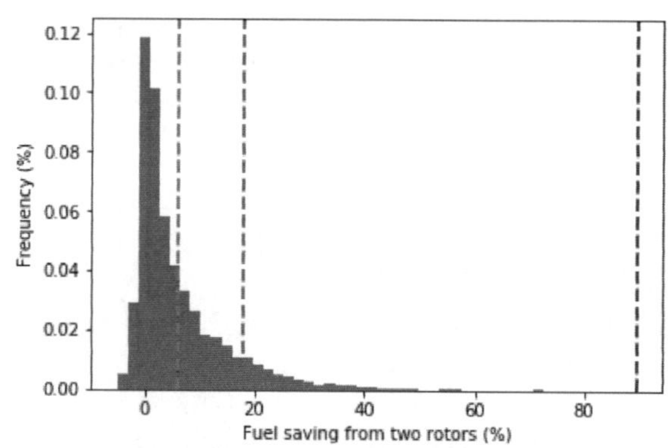

Figure 14: Histogram showing the frequency of fuel savings (in % of fuel saved) from double Flettners for all combined routes on the Baltic Sea. The average (red), 90th percentile (green) and maximum (blue) of the distribution are shown.

The maximum fuel saving, however, shows a marked difference to this pattern, with the double Flettner providing a maximum fuel saving of 89.6%, compared to 34.2% for the Single Flettner case. These two maximum fuel savings were experienced on exactly the same simulation date and route, which saw both very large wind speeds and beneficial wind directions.

It can also be seen from the distribution of fuel savings in Figure 14, that a small proportion of the fuel savings are negative. For these cases, the vessel with rotors installed is consuming more fuel than if there were no rotors installed at all. This is expected, due to the added resistance of the rotors in unfavourable weather conditions.

5.2 d) Monthly variation

Monthly-averaged fuel savings were calculated for all four combined routes, to assess the monthly variation in fuel savings throughout the full year of 2015. The results from the double Flettner case can be seen in Figure 15. Average fuel savings for each month vary, with values ranging between a minimum of 2.06% in June to a maximum of 9.61% in December. The monthly variation in average fuel savings for the Single Flettner scenario is not presented here, but shows a similar distribution of results.

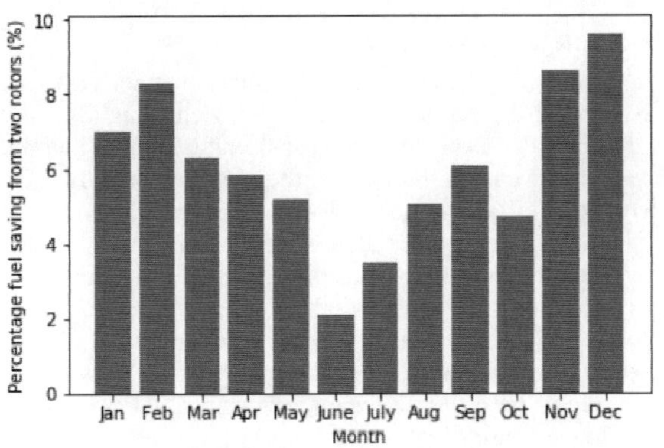

Figure 15: Monthly-averaged fuel savings (in % of fuel saved) obtained from double Flettners for all combined routes on the Baltic Sea.

The results reported in Figure 15 show greater fuel savings in the late autumn and winter months, when winds are stronger, with lower fuel savings in the summer months, especially June and July. However, as these simulations have been conducted for one year only, these conclusions are not statistically valid, and further work is necessary to confirm a relationship between fuel savings and months in the Baltic Sea. However, these initial findings agree with seasonal variations found in other papers for wind-assisted vessels (Smith et al. (2013)).

5.3 ECONOMIC EVALUATION

The price 520€ is highlighted with black dots since this is the current BW0.1S price in Rotterdam. (590USD with 1.14 USD/€ as of 18/01/2019). The variation in price over the last three months was between 500€ and 650€.

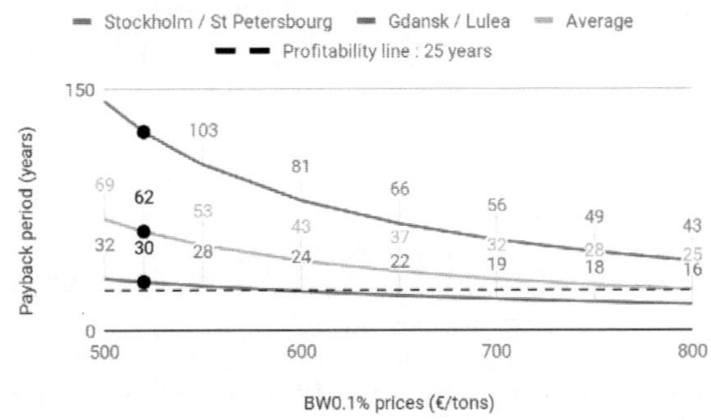

Figure 16: Payback period in Baltic sea under the Single Flettner scenario

The payback period for both cases is presented in Figure 16 and Figure 17 above, with results displayed for Stockholm - St. Petersburg, for Gdansk - Lulea, and as an average for all simulations (indicative for operation in the Baltic region). The installation of WASP technology on a ship operating in the Baltic region will become profitable (meaning with a payback period of less than the assumed 300 months/ 25 years) once the BW0.1S price rises above 800 € / tons for both the Single and Double Flettner scenarios.

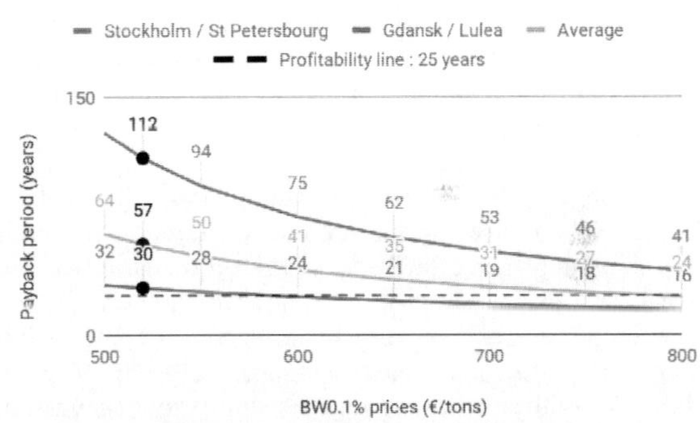

Figure 17: Payback period in Baltic sea under the Double Flettner scenario

The discrepancy in fuel savings between the two routes that was observed in Section 5.2 is evident in the economic performance. A ship operating on the Gdansk - Lulea route becomes profitable when BW0.1S price reaches 586 € / tons, which is within the range of fuel price fluctuations in recent months. The poor savings obtained on the Stockholm - St. Petersburg have a direct consequence for the payback period, which is nearly double the average.

In the definition of the savings function, several assumptions were made due to the absence of available information. An assumption was to neglect the 0.9% reduction in deadweight due to the weight of the rotor and rotor foundation. The change in deadweight may be compensated by the reduced bunkering requirement of wind-assist vessels, where especially for longer voyages, the weight of fuel savings may exceed the rotor weight. As a final observation, the reduced bunkering requirement may lead to financial benefits resulting from a reduced hedging activity against fuel price variations. Further analysis is needed here.

6. CONCLUSIONS

The performance of a wind-assist concept, a 5,150 DWT general cargo vessel, with both one and two Flettner rotors has been assessed using a comprehensive vessel model and routing algorithm on four routes on the Baltic Sea. The average fuel savings for the installation of a Single Flettner is 2.99% and the average fuel savings for the Double Flettner case is 6.11%.

6.1 VESSEL MODELLING

The vessel modelling presented here is the combination of three main modules representing the work of individual authors in each discipline : in aerodynamics, hydromechanics and marine engineering. The description of the vessel modelling has been necessarily brief due to the scope of work presented. A number of design and engineering effects have been highlighted.

From an aerodynamic perspective, the most remarkable result is that the aerodynamic interaction effects occurring between the two installed Flettner rotors decrease the amount of wind-generated thrust that would help reduce the fuel consumption of the ship.

The course-keeping ability of the ship is severely compromised by a destabilizing yaw moment. It is possible to compensate with the rudder to some extent, however the maneuverability of the ship will eventually be impacted. In absence of classification society guidelines, a limit of 10 degrees was adopted in the present modelling. This limit for rudder angles is effectively a limit on the aerodynamic thrust, as can be observed in Figure 11. In fact, for the double Flettner case the vessel is obliged to depower both rotors on upwind courses (see Figure 9, for $30° < TWA < 60°$), indicating that the ship is overcanvassed. The yaw balance of the ship severely limits the operating points for double rotor case. This is a key design concern.

Finally, the engine selection for wind assist cases must be made with care considering the variation of operating point within (or without) the engine envelope, as shown in Figure 13. For a main propulsor arrangement with a fixed pitch propeller, as considered in the present case, a wide envelope is desirable.

6.2 WEATHER ROUTING

The study shows that there is a clear variation in average fuel savings depending on the route traversed in the Baltic Sea, with the highest savings obtained for the route between Lulea and Gdansk. A wide range of savings is also seen for each particular route, with the double Flettner case providing fuel savings as high as 89.6% for one specific simulated voyage, with particularly advantageous weather conditions seen on that date. Initial calculations also demonstrate a variation in fuel savings depending on the month travelled in, with more savings obtained between November and February due to the stronger wind conditions experienced in these months, but further work is needed to confirm the statistical validity of these results. This study, therefore, demonstrates that the fuel savings obtained from a Flettner rotor vary between different routes, and also on the same route over time in the Baltic Sea, which is seen as one of the key challenges to wind-assisted vessels (Nelissen et al., 2016, Rehmatulla et al., 2017).

6.3 ECONOMIC ASSESSMENT

The direct fuel cost savings for operation across the whole Baltic region, with light and variable wind conditions, are not currently sufficient to justify the installation of the Flettner rotors on a 5,150 DWT general cargo vessel operating at 10 knots, when solely incorporating an operational perspective. Under the current BW0.1S prices, the payback period for a ship operating throughout the Baltic region is 57 years for the double Flettner case and 62 years for the single Flettner case, which is significantly more than the expected 25 year-lifetime of the Flettner rotors. However, breaking the assessment of the technology down into individual routes provides a different perspective. If the vessel were restricted to a North-South route (avoiding the Eastern Baltic), then the investment becomes profitable when BW0.1S price reaches 586 € / ton. This is within the range of fuel oil fuel price fluctuations in recent months. Key variables will also have an impact on these results, such as vessel size, speed and Flettner rotor dimensions.

It is important to note that the inclusion of both regulatory- and market-based drivers will also be key factors when considering the suitability of Flettner rotors. For example, the investment in Flettner rotors will prove increasingly beneficial by the favourable impact that installed wind assist technologies have on the vessel EEDI. Wind-assist technologies can contribute to the EEDI rating of the new-build vessel under the category "Innovative energy efficiency technologies", described in the annex of (IMO, 2013). A discussion of the impact of wind assist available effective power on the EEDI calculation will be included in the next presentation of this model. From a market-based perspective, wind-assist trade can be seen as a promising option for investment in the near term, as the push for transparency in shipping broadens to include more types of supply chains.

Future work on wind-assisted vessels in areas of sea with weather conditions similar to those in the Baltic should focus on the incorporation of technologies and practices that enhance the fuel savings of the vessel. For example, combining the technology with route optimisation could increase the profitability of the vessel, by improving the wind conditions experienced on the route to further increase fuel savings. Speed optimisation could also be employed to produce similar results.

Finally, it is understood that the performance of wind assist depends on the prevailing winds along the chosen route. The following presentation of this modelling will include results for a deep-water freighter on a trans-

Atlantic route, where the vessel will enjoy favorable prevailing winds. From an operational perspective, this has the potential of increasing fuel savings resulting in a much improved payback period.

6.4 ONGOING WORK

The aerodynamic model that is currently under development will allow to quickly assess several different positions of the Flettner rotors on the ship's deck in order to find the most favourable ones for the specific case analyzed.

The management of flow separation along the bilges of the hull promises to mitigate the strong 'destabilising' yaw moment that is a consequence of lateral force production by commercial vessel hulls. The sailing performance of commercial hull-types fitted with diverse appendage configurations was investigated in a recent towing tank experiment and will be presented in an upcoming publication.

The next iteration of the engine modelling will include an approach for controllable pitch propellers, a propeller type that is designed to accommodate variable operating points and thought to be well suited to wind assist vessels. Finally, the vessel modelling is to be rebuilt to accommodate ship speed as a variable, enabling more elaborate route optimisation.

Future routing work will involve the incorporation of route optimisation and speed optimisation, where it will be possible for the vessel to deviate from the GCR to look for advantageous weather conditions to further reduce fuel consumption. This, along with the inclusion of varying forecasted weather forecast data for weather conditions over time, will be developed into the model for future studies.

Finally, direct contact with the client will allow for a more nuanced economic assessment. In particular, detailed information is needed about the operating profile, the consequences of financial hedging strategies for fuel price, and finally, possible marketing strategies of the client with respect to green supply chains.

6.5 COLLABORATION IN FUTURE WORK

Wind-assisted ship propulsion is a dedicated research theme at the Ship Hydromechanics section of Delft University of Technology. This research program is conducted in close cooperation with partners across the Dutch maritime sector. Relevant research topics encompass a wide range of disciplines, extending beyond physical modelling to include logistics and economics. The further development of this promising technology, and its eventual implementation, will be made possible by these academic, research and industry partners working in concert. The Ship Hydromechanics section is actively pursuing further collaboration in the form of shared publications and shared applications for national and international funding instruments.

7. ACKNOWLEDGEMENTS

The Authors would like to thank Jochem de Jong and Willem Kroon at DAMEN Shipyards, and Tuomas Riski of Norsepower for their assistance with this work.

Part of this research is funded by The Engineering and Physical Sciences Research Council (EPSRC). Additionally, the Authors would like to thank John Broderick, Alice Larkin and Paul Gilbert for their PhD supervision and stimulating discussions on this work. The Authors would also like to thank the Vaderland Fonds and the Bijlboegfonds for their support.

8. REFERENCES

ACCIARO, M, HOFFMAN, P.N., EIDE, M.S. (2013) *The energy efficiency gamp in maritime transport.* Journal of Shipping Ocean Eng. 3(10): 1-10, 2013.

BARAI, Z., LONGVA, T. (2011). Estimated CO2 Emissions Reduction From Introduction of Mandatory Technical and Operational Energy Efficiency Measures for Ships. IMO MEPC 63/INF 2, 2011.

BORDOGNA, G., MUGGIASCA, S., GIAPPINO, S., BELLOLI., M., KEUNING, J.A., HUIJSMANS, R.H.M., VAN 'T VEER, R. (2018) *Experiments on a Flettner rotor at high Reynolds numbers.* Manuscript submitted for publication

BOWS-LARKIN, A., MANDER, S., GILBERT, P., TRAUT, M., WALSH, C. and ANDERSON, K. (2014) *High Seas, High Stakes.* High Seas Final Report. Tyndall Centre for Climate Change Research

BUHAUG, Ø., CORBETT, J.J., ENDRESEN, Ø., EYRING, V., FABER, J., HANAYAMA, S., LEE, D.S., LEE, D., LINDSTAD, H., MARKOWSKA, A.Z., MJELDE, A., NELISSEN, D., NILSEN, J., PÅLSSON, C., WINEBRAKE, J.J., WU, W. and YOSHIDA, K. *Second IMO GHG Study* (2009), International Maritime Organization (IMO) London, UK, April 2009

CHEN, C., HSIEH, T., HSU, T. (2015) *A novel approach to solve the great circle track based on rotation transformation.* Journal of Marine Science and Technology. 23 13-20 DOI 10.6119/JMST-013-1212-1

DIJKSTRA, E. (1959) *A note on two problems in connection with graphs.* Numerische Mathematik, 1(1) 269-271 DOI 10.1007/bf01386390

EGGERS, R. (2016) *Operational Performance of Wind Assisted Ships,"* 10th Symposium on High- Performance Marine Vehicles (HIPER), Cortona, Italy, 2016.

FABER, J., MARKOWSKA, A., NELISSEN, D., DAVIDSON, M., EYRING, V., CIONNI, I., ET AL

(2009) *Technical support for European action to reducing Greenhouse Gas Emissions from international maritime transport*, CE Delft. 2009.

FUJIWARA, T., HEARN, G., KITAMURA, F., UENO, M., MINAMI, Y. (2005). *Stead sailing performance of a hybrid-sail assisted bulk carrier*. Journal of Marine Science and Technology, pp. 131-146, 2005.

FUJIWARA, T., UENO, M. and IKEDA, Y. (2005) *A new estimation method of wind forces and moments acting on ships on the basis of physical component models.* Journal JASNOE, Vol. 2

GERRITSMA, J., BEUKELMAN, W. (1972) *"Analysis of the Resistance Increase in Waves of a Fast Cargo Ship".* International Shipbuilding Progress Vol. 19 Issue 217.

GOOGLE MAPS. (2019) *Google Maps.* [online] Available at: https://www.google.com/maps/d/embed?mid=15uIgdviES-EURVyK11fkCThF8hwG1-ac

IMO, 2015 *EEDI Database – Review of Status of Technological Development – Note by the Secretariat*. IMO MEPC 68/INF.13, 2015.

KANG, J. and HUSTVEDT, G. (2014) *Building Trust Between Consumers and Corporations: The Role of Consumer Perceptions of Transparency and Social Responsibility.* Journal of Business Ethics 125 (2):1-13, 2014.

LITTLER, J. *"What's wrong with ethical consumption?"* In: Lewis, T and Potter, E (Eds.), Ethical Consumption: A Critical Introduction. (pp. 27-39). Routledge. (2011).

MATSUMOTO, N., INOUE, M., SUDO, M. (1982). *2nd International Conference on Stability of Ships and Ocean Vehicles,* STAB '82, Tokyo, pp 451–464

MOL, A., (2013) *Transparency and value chain sustainability.* Journal of Cleaner Production. 107. Nov. 12 2013

NAAIJEN, P., SHI, W. and KHERIAN, J. (2010) *Assessing the fuel savings by using auxiliary wind propulsion from traction kites.* RINA, Royal Institution of Naval Architects - Ship Design and Operation for Environmental Sustainability - Papers. 27-37.

NELISSEN, D., TRAUT, M., KÖHLER, J., MAO, W., FABER, J., and AHDOUR, S., (2016) *Study on the analysis of market potentials and market barriers for wind propulsion technologies for ships.* DOI 10.2834/68747

NIELSON SURVEY www.nielsen.com, Oct 10, 2015

PULIZZI, J. (2012) *The Rise of Storytelling as the New Marketing.* Publishing Research Quarterly 28(2), 2012

REHMATULLA, N., PARKER, S., SMITH, T. and STULGIS, V. (2017) *Wind technologies: Opportunities and barriers to a low carbon shipping industry.* Marine Policy 75 217-226, DOI 10.1016/j.marpol.2015.12.021

SAMITAS, A. and TSAKALOS, I. (2010) *International Journal of Financial Markets and Derivatives.* Vol. 1, issue 2, 196-212, 2010

SHIP & BUNKER. (2019 *Ship & Bunker* [online] Available at: https://shipandbunker.com/prices/emea/nwe/nl-rtm-rotterdam#ULSFO

SMITH, T, NEWTON, P, WINN, G, GRECH LA ROSA, A. (2013) *"Analysis techniques for evaluating the fuel savings associated with wind assistance.* Low Carbon Shipping Conference, London 2013.

STRATIOTIS, E. (2015) *Fuel costs in ocean shipping.* http://www.morethanshipping.com, (Accessed 22nd January 2018)

TRAUT, M., GILBERT, P., WALSH, C., BOWS, A., FILIPPONE, A., STANSBY, P. and WOOD, R. (2014) *Propulsive power contribution of a kite and a Flettner rotor on selected shipping routes.* Applied Energy, 113 362-372, DOI 10.1016/j.apenergy.2013.07.026

TRAUT, M., LARKIN, A., ANDERSON, K., MCGLADE, C., SHARMINA, M. and SMITH, T. (2018) *CO2 abatement goals for international shipping.* Climate Policy 0(0) 1-10 DOI 10.1080/14693062.2018.1461059

VAN DER KOLK, N.J., AKKERMAN, I., KEUNING, J.A., HUIJSMANS, R.H.M. (2019) *"Part 1: Experimental Validation of a RANS-CFD Methodology for the Hydrodynamics of Wind-Assisted Ships Operating at Leeway Angles"* (2019) Ocean Engineering. Accepted for publication.

VAN DER KOLK, N.J., AKKERMAN, I., KEUNING, J.A., HUIJSMANS, R.H.M. (2019) *"Part 2: Simulation Methodology and Numerical Uncertainty for RANS-CFD for the Hydrodynamics of Wind-Assisted Ships Operating at Leeway Angles".* (2019) Manuscript submitted for publication

VAN WALREE, F. and WILLEMSEN, E. (1988) *Wind loads on offshore structures.* BOSS-88 Conference, Trondheim 1988

WEI, S., GRIMMELIUS, H.T. and STAPERSMA, D. (2010) *Analysis of ship propulsion system behaviour and the impact on fuel consumption.* International Shipbuilding Progress 57 (2010) 35–64 35, DOI 10.3233/ISP-2010-0062

WOODSIDE, A.G., SOOD, S. and MILLER, K. (2008) *When consumers and brands talk: Storytelling theory and research in psychology and marketing.* Wiley Periodicals, 2008

AUTHORS' CONTACT DETAILS

Chris Watts
Babcock International Group
Chris.Watts@babcockinternational.com

Fabrizio Cadenaro
Lloyd's Register EMEA
Fabrizio.Cadenaro@lr.org

Ed Fort
Lloyd's Register EMEA
Ed.Fort@lr.org

Liam Blackmore
Lloyd's Register EMEA
Liam.Blackmore@lr.org

John Buckingham
BMT
john.buckingham@bmtglobal.com

David Pearson
BMT
david.pearson@bmtglobal.com

Maja Perčić
University of Zagreb
Maja.Percic@fsb.hr

Ivica Ančić
University of Zagreb
Ivica.Ancic@fsb.hr

Nikola Vladimir
University of Zagreb
Nikola.Vladimir@fsb.hr

Aykut I Ölçer
World Maritime University
AIO@wmu.se

Dimitrios Dalaklis
World Maritime University
dd@wmu.se

Fabio Ballini
World Maritime University
fb@wmu.se

Eko Maja Priyanto
Biro Klasifikasi Indonesia
S17071@alumni.wmu.se

Victor Bolbot
University of Strathclyde
victor.bolbot@strath.ac.uk

Gerasimos Theotokatos
University of Strathclyde
gerasimos.theotokatos@strath.ac.uk

Evangelos Boulougouris
University of Strathclyde
evangelos.boulougouris@strath.ac.uk

Dracos Vassalos
University of Strathclyde
d.vassalos@na-me.ac.uk

Jon Nation
Wärtsilä Marine Solutions
jon.nation@wartsila.com

Juho Rekola
Steerprop Ltd
juho.rekola@steerprop.com

Martin Andtfolk
Steerprop Ltd
terhi.salomaa@steerprop.com

Giovanni Bordogna
TUDelft
G.Bordogna@tudelft.nl

Paul Desprairies
TUDelft
paul.desprairies@gmail.com

Wandifa Saidyleigh
Gambia Maritime Administration
S17096@alumni.wmu.se

Ville Paakkari
Norsepower Oy Ltd
ville.paakkari@norsepower.com

Chester Lewis
E4tech Ltd
chester.lewis@e4tech.com

Kjell Storelid
Wärtsilä Marine Solutions
kjell.storelid@wartsila.com

Nico van der Kolk
TUDelft
N.J.vanderKolk@tudelft.nl

James Mason
University of Manchester
james.mason@manchester.ac.uk

AusilioBauen
E4tech Ltd
ausilio.bauen@e4tech.com

Raphael Baumler
World Maritime University
rb@wmu.se

The Royal Institution of Naval Architects
8-9 Northumberland Street
London, WC2N 5DA
United Kingdom

ISBN 978-1-5108-9541-6

CURRAN ASSOCIATES INC.
proceedings.com